AF845344

Development of a Novel Method for the Analysis of Haloacetic Acids in Drinking Water

The mission of the Awwa Research Foundation (AwwaRF) is to advance the science of water to improve the quality of life. Funded primarily through annual subscription payments from over 1,000 utilities, consulting firms, and manufacturers in North America and abroad, AwwaRF sponsors research on all aspects of drinking water, including supply and resources, treatment, monitoring and analysis, distribution, management, and health effects.

From its headquarters in Denver, Colorado, the AwwaRF staff directs and supports the efforts of over 800 volunteers, who are the heart of the research program. These volunteers, serving on various boards and committees, use their expertise to select and monitor research studies to benefit the entire drinking water community.

Research findings are disseminated through a number of technology transfer activities, including research reports, conferences, videotape summaries, and periodicals.

Development of a Novel Method for the Analysis of Haloacetic Acids in Drinking Water

Prepared by:
Li Zhang and **Raymond M. Hozalski**
Department of Civil Engineering, University of Minnesota
500 Pillsbury Dr. SE, Minneapolis, MN 55455

Sponsored by:
Awwa Research Foundation
6666 West Quincy Avenue, Denver, CO 80235-3098

Published by:

Awwa Research Foundation
Advancing the Science of Water®

IWA Publishing

DISCLAIMER

This study was funded by the Awwa Research Foundation (AwwaRF). AwwaRF assumes no responsibility for the content of the research study reported in this publication or for the opinions or statements of fact expressed in the report. The mention of trade names for commercial products does not represent or imply the approval or endorsement of AwwaRF. This report is presented solely for informational purposes.

Published by IWA Publishing, Alliance House, 12 Caxton Street, London SW1H 0QS, UK

Telephone: +44 (0) 20 7654 5500; Fax: +44 (0) 20 7654 5555; Email: publications@iwap.co.uk;
Web: **www.iwapublishing.com**

AwwaRF report number 91013F. Originally published by AwwaRF for its subscribers in 2004.
IWA Publishing version published 2005.

Copyright © 2004
by
Awwa Research Foundation

ISBN 1 84339 906 7

CONTENTS

TABLES

FIGURES

FOREWORD

The Awwa Research Foundation is a nonprofit corporation that is dedicated to the implementation of a research effort to help utilities respond to regulatory requirements and traditional high-priority concerns of the industry. The research agenda is developed through a process of consultation with subscribers and drinking water professionals. Under the umbrella of a Strategic Research Plan, the Research Advisory Council prioritizes the suggested projects based upon current and future needs, applicability, and past work; the recommendations are forwarded to the Board of Trustees for final selection. The foundation also sponsors research projects through an unsolicited proposal process; the Collaborative Research, Research Applications, and Tailored Collaboration programs; and various joint research efforts with organizations such as the U.S. Environmental Protection Agency, the U.S. Bureau of Reclamation, and the Association of California Water Agencies.

This publication is a result of one of these sponsored studies, and it is hoped that its findings will be applied in communities throughout the world. The following report serves not only as a means of communicating the results of the water industry's centralized research program but also as a tool to enlist the further support of the nonmember utilities and individuals.

Projects are managed closely from their inception to the final report by the foundation's staff and large cadre of volunteers who willingly contribute their time and expertise. The foundation serves a planning and management function and awards contracts to other institutions such as water utilities, universities, and engineering firms. The funding for this research effort comes primarily from the Subscription Program, through which water utilities subscribe to the research program and make an annual payment proportionate to the volume of water they deliver and consultants and manufacturers subscribe based on their annual billings. The program offers a cost-effective and fair method for funding research in the public interest.

A broad spectrum of water supply issues is addressed by the foundation's research agenda: resources, treatment and operations, distribution and storage, water quality and analysis, toxicology, economics, and management. The ultimate purpose of the coordinated effort is to assist water suppliers to provide the highest possible quality of water economically and reliably. The true benefits are realized when the results are implemented at the utility level. The foundation's trustees are pleased to offer this publication as a contribution toward that end.

Walter J. Bishop
Chair, Board of Trustees
Awwa Research Foundation

James F. Manwaring, P.E.
Executive Director
Awwa Research Foundation

ACKNOWLEDGMENTS

The authors of this report are indebted to Paul Capel of the U.S. Geological Survey and Adjunct Professor at the University of Minnesota for advice and help with solid phase extraction. In addition, the advice provided by the Project Advisory Committee members (Susan Andrews of the University of Waterloo, Jeff Czarnecki of Greenville Water System, and Viorica Lopez-Avila of Midwest Research Institute) and the help from our Awwa Research Foundation project officer, Djanette Khiari, are sincerely appreciated. Finally, the authors acknowledge Saint Paul Regional Water Services of Saint Paul, Minnesota, for providing water samples and data used for this project.

EXECUTIVE SUMMARY

BACKGROUND

Three U.S. Environmental Protection Agency (EPA)-approved methods are currently available for analysis of haloacetic acids (HAAs) in drinking water, Standard Method 6251B, EPA method 552.1, and EPA method 552.2 (APHA, AWWA, and WEF 1998; USEPA 1992; USEPA 1995). In these methods, HAAs are extracted from the water samples using either methyl tert-butyl ether (MTBE) (6251 B and 552.2) or anion exchange columns (552.1), converted into their corresponding methyl esters using either diazomethane (6251 B) or acidic methanol (522.1 and 552.2), and analyzed by gas chromatography (GC) with electron capture detection. Other methods involving GC separation coupled with mass spectrometry detection have been reported in the literature (Xie 2001, Martinez et al. 1998b, Sarrion et al. 2000). These methods also require a tedious sample derivatization step due to the low volatility and high polarity of HAAs. No derivatization step is necessary for methods using reversed-phase ion interaction chromatography or high performance liquid chromatography, but the detection limits for these method are significantly greater than typical drinking water concentrations which are in the low mg/L range (Sarzanini et al. 1999, Carrero and Rusling 1999). Lopez-Avila (2000) developed a method involving serial extractions and HAA analysis by ion chromatography with a conductivity detector, but the authors reported a problematic sulfate interference issue. Methods using electrospray ionization mass spectrometry or high field asymmetric waveform ion mobility spectrometry-mass spectrometry provide sufficiently low HAA detection limits, but these instruments are expensive and not widely available (Ells et al. 2000, Magnuson and Kelty 2000, Loos and Barcelo 2001, Roehl et al. 2002).

Capillary electrophoresis (CE) is an extremely effective technique for separating charged solutes in aqueous solutions. In the last 3 to 4 years, several investigations concerning HAA analysis by CE have been reported in the literature (Kim et al. 2001, Xie and Romano 1997, Xie et al. 1999, Ahrer and Buchberger 1999, Martinez et al. 1998a, Martinez et al. 1998c, Martinez et al. 1999). Phosphate or 2,6-naphthalenedicarboxylic acid dipotassium (NDC) served as the background electrolyte (BGE) in these studies. In general, the CE analysis times were relatively short (less than 10 minutes) and the HAA peaks were well separated. Nevertheless, one concern regarding the use of NDC is an unstable baseline, which commonly occurs with indirect detection methods (Martinez et al. 1998c; Kim et al. 2001). The reported method detection limits were in the low μg/L range for HAAs. Nevertheless, a sample preconcentration step was needed to enrich HAAs prior to separation and detection due to the poor sensitivity resulting from the short path length of the light absorbance detector systems used in most CE analyzers (50 – 100 μm) and the weak light absorbing characteristics of HAAs. Several researchers employed liquid-liquid extraction for HAA preconcentration (Kim et al. 2001, Xie and Romano 1997, Xie et al. 1999, Ahrer and Buchberger 1999). Martinez et al. (1998a, 1998c, 1999) investigated two off-capillary preconcentration techniques, liquid-liquid extraction and solid phase extraction (SPE), and one on-capillary preconcentration technique termed electrokinetic injection followed CE separation with indirect detection. Recoveries with liquid-liquid extraction using MTBE ranged from 60 to 80% (Martinez et al. 1998c). SPE using LiChrolut EN (a high crosslinked styrene-divinylbenzene copolymer) was effective for HAA concentration (Martinez et al. 1998a). Electrokinetic injection significantly increased method sensitivity, however, sample clean up using SPE was required prior to injection (Martinez et al. 1999). More work is needed to study and optimize HAA preconcentration processes for use with CE.

RESEARCH OBJECTIVES

The main objective of this research was to develop a new approach for HAA analysis that was accurate, reproducible, sufficiently sensitive, and relatively safe and simple to perform. This method involved the use a CE ion analyzer for separation and detection of HAAs. Preliminary results from our laboratory using distilled and deionized water spiked with HAAs indicated that this approach was capable of separating and detecting all nine chlorinated and brominated HAAs. The method detection limit (MDL) for monochloroacetic acid measured in our laboratory was 32 ppb using a bubble cell without any sample pretreatment (i.e. direct aqueous sample injection). Given this unacceptably high MDL, more work was needed to fully develop and test the CE-based method especially with regard to improving method sensitivity. Therefore, the specific objectives of this research were as follows:

1. Evaluate the effects of background electrolyte (BGE) and electroosmotic flow (EOF) modifier on HAA separation and detection by CE.
2. Develop and test methods for sample preconcentration including on-capillary techniques (electrokinetic injection and capillary isotachophoresis (CITP)) and an off-capillary technique (solid-phase extraction (SPE)).
3. Evaluate the effects of dissolved constituents present in typical treated water samples (e.g., natural organic matter) on HAA separation, detection and recovery using the CE-based method.
4. Verify the CE-based method developed in this research by comparing results from the analysis of HAAs in finished water samples (both ambient and spiked) with those obtained using an EPA-approved method (EPA Method 552.2).

APPROACH

The first step in this investigation was to evaluate the effects of BGE and EOF modifier on HAA peak resolution and peak shape. Different combinations of BGE (phosphate, citric acid + lithium hydroxide, or NDC) and EOF modifier (tetradecyltrimethylammonium bromide (TTABr), cetyltrimethylammonium bromide, hexadimethrine bromide, or diethylenetriamine) were tested over a range of pH values. The combination providing a stable baseline, resolution of the nine HAA peaks, and sharp peaks was selected for further method development. Then, calibration curves and instrument detection limits (IDLs) were determined for each HAA using the optimal BGE-EOF combination. Seven replicates of a dilute HAA aqueous standard solution were analyzed to determine the IDL for each HAA. The IDLs were calculated as 3.14 times the standard deviation of the seven replicate analyses according to *Standard Methods* (APHA, AWWA, and WEF 1995).

The effects of capillary design (i.e. straight capillary vs. bubble cell capillary) and injection time for pressure injection on method sensitivity and peak resolution were investigated. In addition, three sample preconcentration techniques were evaluated: two on-capillary techniques (electrokinetic injection and capillary isotachophoresis (CITP)) and off-capillary sample preconcentration using SPE. The linearity of calibration curves and interference from sample matrix constituents were evaluated for the electrokinetic injection approach. For CITP, various combinations of leading and terminating electrolyte were evaluated. LiChrolut EN columns containing highly crossed-linked styrene-divinylbenzene resin from EM Science were selected for SPE preconcentration. The effects of sample pH, sample volume, sample loading rate, rinse water pH, rinse water volume and methanol elution volume on HAA recoveries and concentration factor were investigated to optimize the SPE procedure. Then, the HAA MDLs were determined for the final preconcentration-CE method.

The next step in the method development involved an investigation of the effects of sample matrix constituents (Cl, Br, and natural organic matter) on HAA recoveries using the preconcentration-CE method.

The last step in the method development was method verification. The results obtained from the analysis of HAAs in finished water samples (both ambient and spiked) using the preconcentration-CE method were compared with those obtained using EPA method 552.2.

CONCLUSIONS

1. The optimal BGE-EOF combination for analysis of HAAs was phosphate (25 mM)-TTAOH (1 mM) at pH 5.7. Other combinations were eliminated for a variety of reasons including poor HAA peak resolution, poor peak shapes, or unstable detector baseline.
2. Increasing the sample injection from 200 mbar•s to 600 mbar•s and replacing the straight capillary with a bubble-cell capillary each provided an approximately 3-fold decrease in the HAA IDLs without a detrimental effect on peak resolution.
3. Neither of the on-capillary preconcentration techniques was successful. Electrokinetic injection was disregarded because of the non-linearity of the calibration curves and severe sample matrix interferences. CITP was not successful because the HAA sample peaks were unresolved under all conditions tested.
4. SPE using LiChrolut EN columns was successful at enriching all nine HAAs. The SPE process includes six steps: 1) preconditioning of the column with 30 mL of methanol and 10 mL of pH 2 high performance liquid chromatography (HPLC)-grade water, 2) acidification of the sample to pH 2, 3) loading of the sample (25 mL or 50 mL) at 1 mL/min, 4) column rinsing with 0.5 mL of pH 2 HPLC grade water, 5) elution of HAAs with 2.3 mL of methanol, 6) dilution with 0.2 mL of HPLC grade water and 7) evaporation of methanol prior to analysis.
5. Tribromoacetic acid (TBAA) (200 µg/L) is the recommended surrogate for determining HAA extraction efficiencies. TBAA was selected because it has a similar structure and similar physical-chemical properties as the HAA_5 compounds, it does not interfere with the other HAA peaks during CE analysis, and it is rarely detected in treated drinking water.
6. Recoveries were in the range of 75% to 99% for all HAAs except CAA (63%) when 50 mL of sample was passed through the SPE column. The recovery for CAA increased to 80% when only 25 mL of sample was processed but the HAA concentration factors decreased by a factor of two. HAAs sorbed onto SPE columns are stable except for TBAA whose recovery appeared to decrease over a 5-day column storage period.
7. The final SPE-CE method for HAA analysis includes a SPE preconcentration step followed by CE analysis. The SPE procedure was described in conclusion (4) above.

Pressure injection (50 mbar for 12 s) is used to load the samples onto the capillary. A bubble cell capillary (75 μm ID × 56 cm effective length) is used for HAA separation and the detector wavelength is 191 nm. The time to process each sample is approximately 5 hours but multiple samples can be processed at the same time with the use of a multi-port SPE vacuum manifold.

8. The SPE-CE method can be used to analyze for all HAAs except chlorodibromoacetic acid, which consistently comigrated with a contamination peak, and TBAA because it serves as the surrogate. The HAA MDLs for the SPE-CE method ranged from 1.7 to 2.5 μg/L for all HAAs except CAA (5 μg/L).
9. The results from the analysis of HAAs in finished water samples obtained using the SPE-CE method were in good agreement with those obtained using EPA method 552.2. The HAA peaks in the finished water samples were identified using relative migration times, calculated as the migration time of the target compound divided by the migration time of the surrogate (i.e., TBAA).

RECOMMENDATIONS

1. The SPE-CE method developed in this research effort is recommended for those analytical laboratories that have safety concerns regarding HAA methylation, especially the preparation and use of diazomethane. The SPE-CE method is also recommended for those laboratories struggling with HAA methylation efficiencies and recoveries using the EPA-approved methods. The SPE-CE method represents a significant improvement in terms of safety as the chemicals and conditions used are relatively benign. Unfortunately, the need for off-capillary HAA preconcentration by SPE represents a significant time commitment so there is no significant savings in time by switching to the SPE-CE method. Future developments using currently available technology (i.e. robotics) could eventually lead to automation of the SPE preconcentration process if the demand for such a process was sufficent.
2. The relatively low CAA recovery (63%) when 50 mL of sample is preconcentrated by SPE is a concern for quantification of CAA. Thus, two runs may be necessary for the analysis of each sample if the 63% recovery for CAA is deemed unacceptable, one run in which 25 mL of sample is preconcentrated for analysis of CAA, and the other in which 50 mL is preconcentrated for analysis of the remaining HAAs.
3. To achieve sub-μg/L MDLs using the SPE-CE method developed in this research, the use of electrokinetic injection following SPE should be explored further.

CHAPTER 1
INTRODUCTION AND LITERATURE REVIEW

INTRODUCTION

Haloacetic acids (HAAs) are disinfection byproducts (DBPs) formed during the chlorination of water and wastewater (Krasner et al. 1989, Rebhun et al. 1997). HAAs and other DBPs such as trihalomethanes (THMs) form when hypochlorous acid (HOCl) reacts with the organic matter present in these waters. Brominated HAAs can also be produced when HOCl oxidizes bromide to hypobromous acid (HOBr), which then reacts with organic matter (Heller-Grossman et al. 1993). HAAs are typically present at ng/L concentrations in surface waters (Berg et al. 2000, Müller et al. 1996, Scott et al. 2000), µg/L concentrations in treated drinking water (Heller-Grossman et al. 1993, Krasner et al. 1989), and µg/L to mg/L concentrations in treated wastewater (Berg et al. 2000, Rebhun et al. 1997). The occurrence of HAAs in surface waters likely results from industrial and municipal wastewater discharges (Berg et al. 2000, Müller et al. 1996) and from atmospheric deposition (Müller et al. 1996).

There are nine possible chlorinated and brominated HAAs including chloroacetic acid (CAA), bromoacetic acid (BAA), dichloroacetic acid (DCAA), bromochloroacetic acid (BCAA), dibromoacetic acid (DBAA), trichloroacetic acid (TCAA), bromodichloroacetic acid (BDCAA), chlorodibromoacetic acid (CDBAA), and tribromoacetic acid (TBAA). Chloroacetic acids are phytotoxic and CAA and TCAA were used as herbicides until the late 1980s (Berg et al. 2000). Furthermore, all HAAs are suspected carcinogens and pose a potential health risk to consumers of chlorinated drinking water (National Academy of Sciences Safe Drinking Water Committee 1980). Therefore, HAAs are currently regulated by the United States Environmental Protection Agency (USEPA). Under the Stage II Disinfectants/Disinfection By-Products (D/DBP) Rule, the maximum contamination level (MCL) is set at 60 µg/L for the sum of five HAAs (HAA_5 = CAA+DCAA+TCAA+BAA+DBAA) based on peak DBP sites in a water distribution system (USEPA 2000). In order to monitor HAAs concentrations in drinking water, a robust analysis method is needed.

LITERATURE REVIEW

EPA Approved Methods

Three USEPA approved methods are currently available for analysis of HAAs, including Standard Method 6251B, EPA method 552.1, and EPA method 552.2 (APHA, AWWA, and WEF 1998; USEPA 1992; USEPA 1995). Method 6251B involves liquid-liquid extraction of the HAAs from water using methyl tert-butyl ether (MTBE), derivatization of the HAAs to form the corresponding methyl esters using diazomethane, and analysis by gas chromatography with electron capture detection (GC-ECD) (Figure 1.1). Preparation of diazomethane requires 1-methyl-3-nitro-1-nitrosoguanidine (MNNG), which is a potent carcinogen (Windholz et al. 1983). Furthermore, the diazomethane itself presents a potential explosion hazard. In addition, there is relatively low method sensitivity for CAA due to its low MTBE extraction efficiency. Finally, significant time and effort are required to perform the entire method as shown in Figure 1.1, which does not include the preparation of the diazomethane from MNNG.

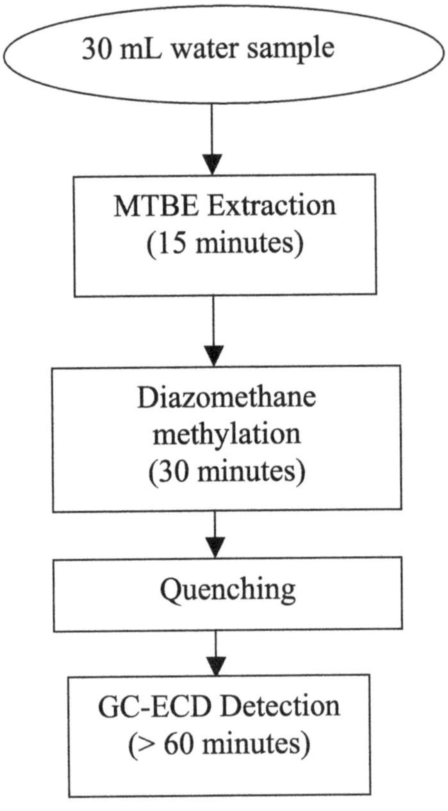

Figure 1.1 Flow diagram for HAA method 6251B (adapted from Xie et al. 1998)

In EPA method 552.1, HAAs are extracted via a solid-phase extraction procedure using anion exchange columns or disks instead of liquid-liquid extraction. In addition, acidic methanol is used in place of diazomethane for HAA methylation (Figure 1.2). Although no diazomethane is used in EPA method 552.1, the method still has some drawbacks including: interferences in drinking water, especially sulfate, that affect the recoveries of some HAAs; lack of reproducibility in the methylation step for some HAAs; and possible GC column damage caused by the acidic methanol methylation step (Lopez-Avila 2000). Finally, significant time and effort are required to perform the entire method (Figure 1.2).

EPA method 552.2 involves liquid-liquid extraction using MTBE, acidic methanol methylation, and analysis by GC-ECD (Figure 1.3). Although improvements were made to method 552.1, such as neutralization of the acidic methanol, there are still some potential problems. Most importantly, the brominated trihaloacetic acids can undergo decarboxylation during methylation and the accuracy of the analytical results for those compounds depends on the sample matrix and the analyst's skill level (Magnuson and Kelty 2000, Urbansky and Magnuson 2002). Furthermore, there is significant time and effort required to perform the entire method (Figure 1.3).

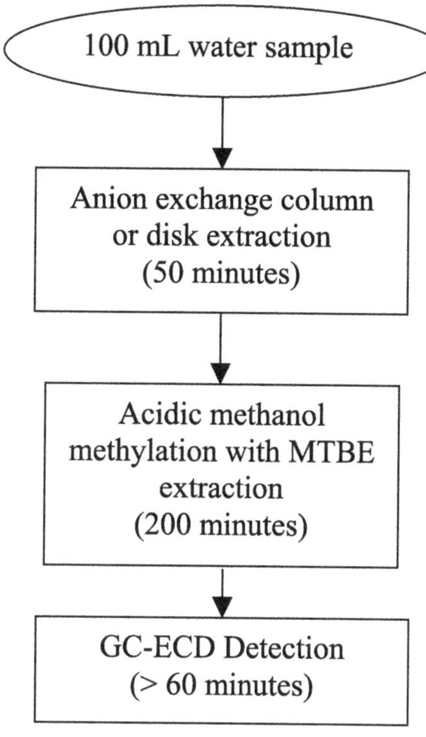

Figure 1.2 Flow diagram for HAA method EPA 552.1

Alternative Methods

Several other HAA analysis methods have been reported in the literature. For example, high pressure liquid chromatography using an electrochemical electrode coated with a surfactant-Nafion film was used to analyze for six bromo- and chloro-acetic acids (Carrero and Rusling 1999). The method avoids the derivatization step, however, the HAA detection limits (120 to 10,000 μg/L) were more than two orders-of-magnitude greater than the EPA-approved GC-ECD methods and much greater than typical drinking water concentrations. Following solid-phase extraction, HAAs were analyzed by ion-pair liquid chromatography-electrospray ionization mass spectrometric detection (Loos and Barcelo 2001). The detection limits for HAAs were in the range of 0.2-1.6 μg/L. The major limitations of this method are the high cost and limited availability of electrospray mass spectrometers.

Kampioti and Stephanou (1999) reported the determination of HAAs using the Grob closed-loop stripping extraction technique and subsequent analysis by GC-ECD. The HAA detection limits were in the range of 45-72 ng/L. Nevertheless, only DCAA and DBAA were analyzed using this method. A simultaneous extraction-derivatization procedure followed by GC-ECD detection was developed by Benanou et al. (1998). The HAA detection limits ranged from 0.1 to 0.2 μg/L; however, TBAA could not be analyzed using this method because it underwent decarboxylation.

Separation of haloacetic acids using reversed-phase ion interaction chromatography with UV detection and suppressed and non-suppressed anion exchange chromatography was investigated by Sarzanini et al. (1999). The detection limits for HAAs (7 to 40 μg/L) were much

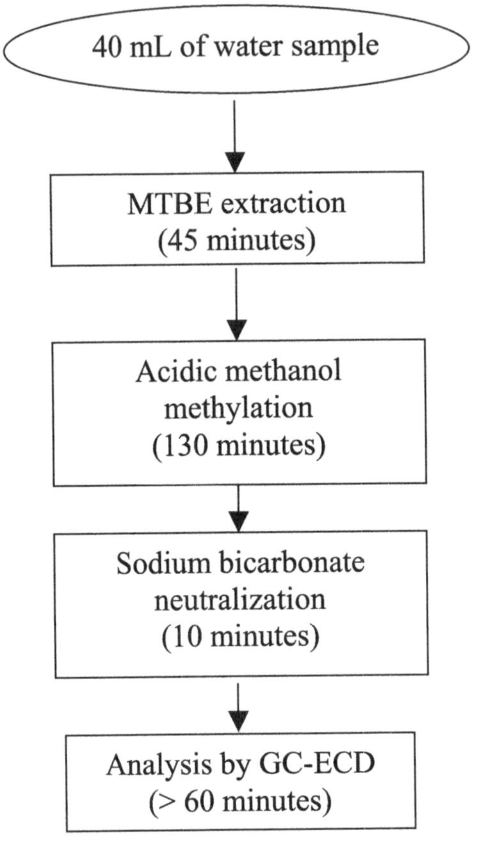

Figure 1.3 Flow diagram for HAA method EPA 552.2

greater than those obtained using the EPA-approved methods. Lopez-Avila (2000) developed a procedure involving microextraction of HAAs with MTBE, back extraction into reagent water, and subsequent analysis by ion chromatography. Nevertheless, the method was deemed unsuitable for routine water quality analysis because of interferences from sulfate and nitrate in drinking water. Roehl et al. (2002) developed an ion chromatography with electrospray mass spectrometric detection method to analyze HAAs in water. The detection limits for the nine haloacetic acids were in the range of 0.2-1.0 µg/L. Again, the major limitations of this method are the high cost and limited availability of electrospray mass spectrometers.

A gas chromatography-mass spectrometry (GC-MS) method was developed by Xie (2001) to analyze for HAAs in water. The sample preparation step is similar to that in EPA method 552.2. Compared to EPA 552.2, this method has several advantages including a cleaner baseline, fewer interfering peaks, and a reduced run time. However, the methylation efficiencies for brominated trihaloacetic acids were low and matrix interferences were reported for chloroacetic acid. Martinez et al. (1998b) developed a gas chromatography-mass spectrometry method to determine HAAs in water. A solid-phase extraction process using strong anion exchange resins was used to enrich the HAAs prior to analysis. The detection limits for HAAs were between 3 µg/L and 20 µg/L, which are significantly greater than those obtained using the EPA-approved methods. Sarrion et al. (2000) developed a procedure involving derivatization of HAAs to their methyl esters with dimethyl sulfate, headspace sampling using solid phase microextraction (SPME), and

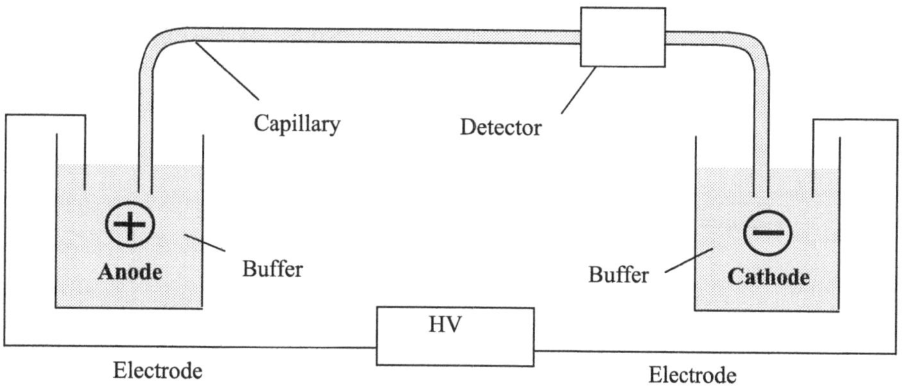

Figure 1.4 Schematic diagram of a capillary electrophoresis system (HV = high voltage)

analysis by gas chromatography-ion trap mass spectrometry (GC-ITMS). Low detection limits in the range of 10 to 450 ng/L were obtained for HAAs. Although very sensitive, this method is relatively complicated and involves a specialized and costly detector.

HAAs were determined by liquid-liquid extraction with MTBE, addition of perfluoroheptanoic acid to the extract to form complexes with HAAs, and subsequent detection of the complexes using electrospray-mass spectrometry. The detection limits for HAAs were below 1 μg/L; however, the standard addition method was needed to quantify the HAAs (Magnuson and Kelty 2000). HAAs were detected at ng/L levels using electrospray ionization, high-field asymmetric waveform ion mobility spectrometry mass spectrometry (ESI-FAIMS-MS) (Ells et al. 2000). Nevertheless, limited availability of ESI-FAIMS-MS will likely restrict the use of this method.

Capillary Electrophoresis for Analysis of HAAs

Capillary electrophoresis (CE) is a relatively new separation technique that has been used primarily for the separation of biomolecules such as proteins and DNA fragments. CE is seeing increasing use in the separation and detection of organic and inorganic ionic species in water. In CE, separation of analytes is achieved by the application of an electric field to a narrow bore fused silica capillary filled with a buffer solution (Figure 1.4). CE systems provide excellent separation efficiencies (number of theoretical plates > 10^5 to 10^6) that are equal to or better than those obtained with chromatographic separation methods (i.e., gas, liquid, or ion chromatography). In addition, analysis times are typically in the range of 2 to 20 minutes, which is much lower than the time for HAA analysis by GC-ECD (> 60 minutes). The basis for separation in capillary electrophoresis is the differences in electrophoretic mobilities of the individual analytes. Sample injection typically occurs near the anode. Most species (including anions) migrate toward the detector (near the cathode) due to electroosmotic flow (EOF) and, in the case of the cations, due to electrostatic attraction as well. Therefore, it is possible to analyze for cations and anions in the same run with cations reaching the detector first, followed by a neutral compound peak, and finally anions, which are attracted to the anode in opposition to the EOF. As the individual analytes in the sample migrate through the buffer-filled capillary, they form discrete zones; hence, the mode of operation is often termed capillary zone electrophoresis (CZE).

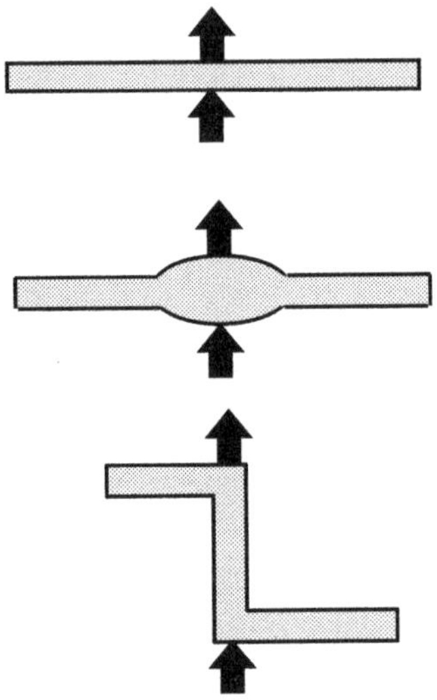

Figure 1.5 Comparison of different CE detector cell designs. The arrows indicate the path of light through the detector cell.

The EOF arises due to the accumulation of cations near the negatively charged fused silica (i.e., SiO_2) capillary wall. The negative charge arises due to the presence of dissociated surface silanol groups ($\equiv SiO^-$) and is pH dependent. As a voltage is applied, the cations migrate towards the cathode and carry the surrounding liquid with them resulting in the EOF. The direction of the EOF can be reversed (i.e., from cathode to anode) by changing the charge on the capillary surface from negative to positive. This is typically accomplished by coating the capillary with a cationic surfactant. In addition, reversing the polarity of the electrodes (i.e., so that the cathode is on the left and the anode on the right in Figure 1) permits the use of the entire capillary length for separation prior to detection. In this way, the system is optimized for analysis of anions by increasing the speed at which these ions travel through the capillary (combined EOF plus electrostatic attraction). The net result is that the anions reach the detector first resulting in a significant reduction in anion analysis time. This EOF reversal approach was used in the analysis of negatively charged HAAs in this research.

For analyte detection, CE systems typically employ either a single wavelength UV light detector or a diode array detector. The detector cell is the capillary itself, which results in a short path length (50 to 100 μm) and consequently, low detection limits. For this reason, recent developments in capillary design have resulted in capillaries with greater pathlengths in comparison to straight capillaries (Figure 1.5). Because of their longer pathlengths, bubble cell capillaries and "z-cell" capillaries can provide approximately 3-fold and 10-fold increases in signal-to-noise ratio (i.e., sensitivity), respectively.

Although CE offers high separation efficiency, the detection limits for analytes such as HAAs in drinking water are typically too high even using the bubble cell or z-cell. Thus, sample preconcentration is often needed to reduce detection limits to acceptable levels. On-line sample stacking or off-line sample preconcentration are two possible approaches for improving method sensitivity. Electrokinetic injection and capillary isotachophoresis (CITP) are two on-line or on-capillary sample stacking techniques. In electrokinetic injection, a voltage is applied to the system with the sample reservoir placed at the capillary inlet. Depending on the polarity of the electrodes, this approach can be used to selectively induce either cations or anions from the sample to enter the capillary.

In CITP, a sample zone is introduced between a leading electrolyte and a terminating electrolyte. The leading electrolyte must have a higher magnitude electrophoretic mobility than that of any of the analyte ions, and the terminating electrolyte must have a mobility lower than that of any of the sample ions. As the run proceeds, sharp zones are formed including the leading electrolyte zone, individual analyte zones, and the terminating electrolyte zone. The fundamental principles of isotachophoresis (ITP) can be described as follows (Krivankova and Bocek 1998):

A. In ITP, the current, which equals the field strength times conductivity, is constant along the capillary such that:

$$I = E_T k_T = E_A k_A = E_L k_L \tag{1.1}$$

where E_T, E_A, E_L are the field strengths in the terminating electrolyte zone, solute zone and leading electrolyte zone, respectively; and k_T, k_A, k_L are the conductivities of the terminating electrolyte, solutes, and leading electrolyte, respectively.

B. At equilibrium, all solutes have the same electrophoretic velocity. Thus

$$v_{iso} = v_T = v_A = v_L \tag{1.2}$$

$$v_{iso} = \mu_T E_T = \mu_A E_A = \mu_L E_L \tag{1.3}$$

where v_{iso} is the isotachophoretic velocity; v_T, v_A, v_L are the velocities of the terminating electrolyte zone, solute zone and leading electrolyte zone, respectively; and μ_T, μ_A, μ_L are the electrophoretic mobilities of the terminating electrolyte, solutes and leading electrolyte, respectively.

C. Suppose an ion of lower mobility (μ_2) enters a zone of ions of higher mobility (μ_1). Since $\mu_2<\mu_1$, it also holds that $k_2<k_1$ and it also follows that $E_2>E_1$. Thus, the ion's migration velocity $v_2 = E_2\mu_2$ decreases to $v_2 = E_1\mu_2$. Consequently, the ion migrates back to its own zone. Thus, sharp boundaries are formed by the step change of electric field strength between the zones.
D. The Kohlrausch equation describes the degree of concentration achieved by ITP:

$$C_L/C_A = [\mu_L/(\mu_L+\mu_R)][(\mu_A+\mu_R)/\mu_A] \tag{1.4}$$

where C_L is the molar concentration of the leading electrolyte, C_A is the analyte molar concentration and the subscript R refers to the counter ion. From this equation, it can be seen that the final concentration of the analyte is proportional to the molarity of the leading electrolyte. Therefore, Equation 5.4 suggests that trace compounds can be concentrated in CITP when high concentrations of leading electrolyte (C_L) are used.

Isotachophoresis followed by CZE can be performed within a single capillary or in a coupled-capillary arrangement. Most of the early work on isotachophoresis was performed using a coupled-capillary arrangement primarily due to the lack of sophisticated capillary

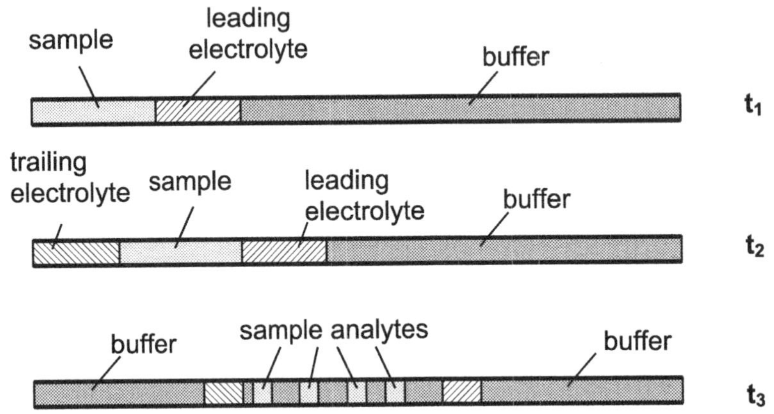

Figure 1.6 Illustration of transient CITP preconcentration in a single capillary arrangement

electrophoresis systems with autosamplers. When isotachophoresis is followed by CZE in a single capillary, then this process is termed "on-capillary transient CITP." In a coupled-capillary system, there are separate capillaries for CITP preconcentration and CZE separation. Transient CITP preconcentration in a single capillary arrangement is illustrated in Figure 1.6 (Foret et al. 1992). First, the column is flushed with buffer. At time t_1, the leading electrolyte is injected into the column followed by the sample. When a suitable terminating electrolyte is applied along with constant current (t_2), the sample can be isotachophoretically preconcentrated. At time t_3, the terminating electrolyte is replaced with the background electrolyte and separation proceeds in the CZE mode.

Kaniansky et al. (1990) evaluated the coupled-capillary CITP-CZE approach for preconcentration and separation of a mixture of nitrophenols and 2,4-dinitrophenyl-labelled amino acids. Low detection limits of 10^{-7} mol/L were achieved using a light absorbance detector. Several researchers have reported significant enhancements in method sensitivity using transient CITP (Altria 1999, Foret et al. 1992). For example, Foret et al. (1992) reported that increases in sensitivity of 50-fold or more were possible with transient CITP compared to CZE alone.

Another approach that may be effective in enhancing CE method sensitivity is off-line sample preconcentration using liquid-liquid extraction or solid phase extraction. Xie and co-workers (Xie and Romano 1997, Xie et al. 1999) developed an MTBE extraction method followed by aqueous back-extraction to improve sensitivity for HAA detection by a capillary electrophoresis ion analyzer. An extraction approach such as this one is likely to provide a one order of magnitude increase in sensitivity. Several other researchers used solid phase extraction (SPE) to preconcentrate HAAs prior to analysis by CE (Martinez et al. 1999, Sarzanini et al. 1999, Martinez et al. 1998a, Loos and Barcelo 2001). Typically, a 1 to 2 order of magnitude increase in sensitivity can be obtained using SPE.

When this research project began, little work in the use of CE for analysis of HAAs had been reported in the literature. However, in the last 3 to 4 years, the results from several parallel investigations have been reported. A procedure involving liquid-liquid extraction followed by indirect UV detection of haloacetic acids was developed by Martinez et al. (1998c). The detection limits for HAAs were in the µg/L range. Martinez et al. (1998a) developed a solid-phase extraction procedure together with CE and indirect UV detection to analyze for HAAs. The detection limits ranged from 2 to 5 µg/L. Martinez et al. (1999) evaluated different separation buffers and used electrokinetic injection for on-line preconcentration of HAAs. The optimum separation buffer contained 4 mM 2,6-naphthalenedicarboxylic acid dipotassium salt (NDC) with 0.5 mM cetyltrimethylammonium bromide (CTABr) as the EOF modifier. The HAA method detection limits (MDLs) were in the low μg/L range. The combination of non-aqueous capillary electrophoresis and mass spectrometry was used to determine HAAs (Ahrer and Buchberger 1999). Liquid-liquid extraction with MTBE was performed prior to sample analysis. The detection limits ranged from 0.3 to 7.6 μg/L; however, poor recoveries were obtained for CAA (33%) and BAA (36%) due to the liquid-liquid extraction process. A procedure involving liquid-liquid extraction with MTBE followed by CE separation and direct UV detection was developed to quantify HAAs in tap water (Kim et al. 2001). HAAs were separated using 25 mM phosphate, 0.5 mM cetyltrimethylammonium chloride (CTACl) at pH 5 and 37°C. The MDLs were in the range of 2 to 5 μg/L but the method was not able to detect TCAA.

SUMMARY

In summary, the current EPA approved methods suffer from interferences from the sample matrix, low recoveries for CAA, and lack of reproducibility for some brominated HAAs. As to the other analysis methods discussed in the literature, the detection limits of HAAs using some methods are much greater than those using the standard method. In addition, tedious and time-consuming liquid-liquid extraction procedures were required in some methods to extract HAAs from sample matrix. Methods involving mass spectrometry for HAA detection (e.g., ESI-FAIMS) can provide low detection limits, however, expensive instruments or detectors are needed. As to the CE methods, some of the work published recently in the literature may overlap with some of the work performed in this research. Thus, we will discuss our results in the context of recent and sometimes overlapping work reported in the literature. However, the research discussed in this report represents a comprehensive study in which the effects of buffer composition, on-line sample concentration, and off-line sample preconcentration on HAA separation and detection by CE were evaluated. In addition, our method was directly compared with a USEPA approved method (Method 552.2) in the analysis of split samples from a local water utility.

CHAPTER 2
PROJECT DESCRIPTION

OBJECTIVES

The main objective of this research was to develop a new approach for HAA analysis that is safe, accurate, reproducible, sufficiently sensitive, and relatively simple to perform. The basis for the method involves the use of a CE ion analyzer for separation and detection of the individual HAAs. The specific objectives of this research were as follows:

1. evaluate the effects of different buffer systems on HAA peak separation and peak shape;
2. evaluate the instrument detection limits (IDLs) for HAAs using a standard or straight capillary and a more sensitive detector cell (i.e., the "bubble cell");
3. develop and test methods for sample concentration including an on-capillary concentration technique (i.e., CITP and electrokinetic injection) and off-line sample preconcentration,
4. evaluate the effect of dissolved constituents present in typical treated water samples (e.g., natural organic matter (NOM)) on HAA separation and detection,
5. apply our CE analysis method to finished water samples obtained from a water utility (St. Paul Regional Water Services, St. Paul, Minn.) and compare the results to those obtained using a USEPA approved extraction-methylation-GC-ECD method.

RESEARCH APPROACH

The research consisted of an analytical chemistry method development project performed in our research laboratory in the Department of Civil Engineering at the University of Minnesota. Initial experiments involved evaluation of the effects of buffer composition, buffer pH and electroosmotic flow (EOF) modifier on HAA peak separation. The method was optimized for HAA detection sensitivity (to achieve sub-µg/L detection limits) using on-line preconcentration and off-line preconcentration techniques. Following development of a suitably sensitive, accurate and reproducible method, we attempted to quantify HAA concentrations spiked into pre-filtered (type GF/F, Whatman Ltd., Maidstone, England) natural water samples obtained from the inlet to the City of St. Paul Water Utility whose primary water source is the Mississippi River. Finally, we compared results for St. Paul finished water samples obtained using our method to those obtained using a USEPA-approved extraction-methylation-gas GC-ECD method.

Evaluation of Different Buffer Systems on HAA Peak Resolution and Peak Shape

Three different buffers were tested: phosphate, citric acid+lithium hydroxide, and NDC. Each buffer was evaluated over a range of pH values. Four different EOF modifiers were used to reverse the EOF: tetradecyltrimethylammonium bromide (TTABr), cetyltrimethylammonium bromide (CTABr), hexadimethrine bromide (HDMBr), and diethylenetriamine (DETA). An experimental matrix was developed to test each buffer and EOF modifier combination and the combination providing good peak resolution and peak shape for HAAs was selected for further evaluation and optimization.

Evaluation of HAA IDLs Using a Straight Capillary and a More Sensitive Detector Cell (i.e., the Bubble Cell Capillary) With Pressure Injection

Neat materials (i.e., pure HAAs) were added to distilled and deionized (DI) water to prepare the HAA spiked samples (Sigma-Aldrich Corp., St. Louis, Missouri). Calibration curves were developed for each HAA over a range of concentrations from 0 to 100 mg/L using both capillary designs and the buffer and EOF modifier combination from objective 1. IDLs for both the straight and bubble cell capillaries were also determined. Seven replicates of a dilute HAA aqueous standard solution were analyzed to determine the instrument detection limit (IDL) for each HAA. The IDLs were calculated as 3.14 times the standard deviation of the seven replicate analyses according to *Standard Methods* (APHA, AWWA, and WEF 1995).

Sample Concentration Including an On-Capillary Concentration Technique (i.e., CITP and Electrokinetic Injection) and Off-Line Sample Preconcentration

An acceptable HAA method must be able to achieve low μg/L MDLs in order to compete with existing methods and, more importantly, to satisfy the analytical need of the water industry. Given the expected HAA detection limits without sample concentration (approximately 200 μg/L), sample concentration prior to separation and detection of the HAAs was required. Therefore, CITP and electrokinetic injection were investigated for on-line preconcentration of HAAs. Improvements in sensitivity, linearity of the calibration curves, and the reproducibility of the methods were evaluated to determine the feasibility of these approaches.

As was discussed in the literature review (Chapter 1), liquid-liquid extraction and solid phase extraction can be used for off-line preconcentration of HAAs. Because liquid-liquid extraction is a tedious process that requires the use of a potentially harmful organic solvent such as MTBE and it produces an extract which may not be suitable for injection into a CE, liquid-liquid extraction was not investigated in this project. In this research, solid phase extraction using LiChrolut EN columns (EM Science, NJ) containing highly crossed-linked styrene-divinylbenzene resin was investigated as a means to concentrate HAAs.

Evaluation of the Effect of Other Solutes on HAA Separation and Detection

Experiments were performed to evaluate the effects of other dissolved species on the separation and detection of HAAs using the CE-based method. Initially, these experiments involved the additions of HAAs to DI water that has been amended with low mg/L concentrations of dissolved salts or NOM. Chloride was used as the representative of anions and Suwanee River humic acid and Suwanee River fulvic acid materials obtained from the International Humic Substances Society was used to represent NOM.

Comparison Between the CE-Based Method and an Extraction-Methylation-GC-ECD Method

Finished water samples (both ambient and spiked samples) obtained from the water distribution system of St. Paul Regional Water Services were analyzed for HAAs and the results were compared to measurements obtained by an outside laboratory using EPA method 552.2. For this phase of the method testing, a water sample was collected and split, with half of the sample sent to the outside laboratory and the other half brought back to our laboratory for analysis using the CE-based method.

CHAPTER 3
EVALUATION OF BUFFER COMPOSITION

INTRODUCTION

The effect of buffer composition on the separation of HAAs was evaluated by testing three different electrolyte or buffer systems: phosphate, citric acid + lithium hydroxide (LiOH), and NDC. Phosphate buffer and NDC buffer were used previously for the separation of HAAs by Xie and Romano (1997) and Martinez et al. (1999), respectively. Citric acid was used for the separation of textile dyes (Oxspring et al. 1995). Each buffer system was evaluated over a range of pH values (Table 3.1). Phosphate, in addition, four different EOF modifiers were evaluated: tetradecyltrimethylammonium hydroxide (TTAOH), cetyltrimethylammonium hydroxide (CTAOH), hexadimethrine hydroxide (HDMOH), and DETA. Those EOF modifiers are commonly used in the analysis of anions by capillary electrophoresis (Galceran et al. 1996, Smyth et al. 1997). A factorial design was used to test the effects of buffer type, buffer concentration, buffer pH, and EOF modifier to determine which buffer/EOF modifier system provides optimized peak resolution and peak shape (Table 3.1).

Table 3.1
Factorial design matrix to test the effect of buffer composition

	Buffer			EOF modifier			
Run	Phosphate (5-12.5 mM) pH 3.4-10	NDC (2-8 mM) pH 3.4-10	Citric acid (50 mM) LiOH (10-70 mM) pH 3.4-10	DETA (5 mM)	HDMOH (5 mg/L)	CTAOH (1.0 mM)	TTAOH (1.0 mM)
1	×			×			
2	×				×		
3	×					×	
4	×						×
5		×		×			
6		×			×		
7		×				×	
8		×					×
9			×	×			
10			×		×		
11			×			×	
12			×				×

Table 3.2
Target compounds used in the study

Compound	Supplier	Purity
TFAA	Aldrich	99+%
CAA	Aldrich	99+%
BAA	Aldrich	99+%
DCAA	Aldrich	99+%
BCAA	SupelCo	100%
DBAA	Aldrich	97%
TCAA	Aldrich	99+%
BDCAA	SupelCo	100%
CDBAA	SupelCo	99+%
TBAA	Aldrich	99+%

MATERIALS AND METHODS

Preparation of HAA Working Solution

The neat materials listed in Table 3.2 were dissolved in distilled and deionized water to prepare individual 1000 mg/L aqueous stock solutions. Some of the HAAs (e.g., CAA, BAA, DBAA, TCAA, BCAA, and BDCAA) are hygroscopic solids and therefore difficult to weigh out accurately. Thus, the concentrations of those stock solutions were approximate. Quantitatively accurate standard solutions were prepared for method calibration as described in Chapter 4. Working solution of 10 mg/L HAAs and 100 mg/L trifluoroacetic acid (TFAA) (as an internal standard) or 10 mg/L HAAs and 10 mg/L TFAA was prepared by combining aliquots of the aqueous stock solutions and diluting with DI water.

Buffer Preparation

Stock solutions of 32 mM NDC, 0.6 M LiOH, 0.12 M citric acid, 50 mM sodium phosphate dibasic ($Na_2HPO_4 \cdot 7H_2O$), 50 mM sodium phosphate monobasic ($NaH_2PO_4 \cdot H_2O$) were prepared from neat materials. NDC, LiOH, and citric acid were obtained from Sigma-Aldrich. $Na_2HPO_4 \cdot 7H_2O$ was obtained from EM Science and $NaH_2PO_4 \cdot H_2O$ was obtained from Fisher Scientific. Stock solutions of 100 mM TTABr, 10 mM CTABr, 20 mg/L HDMBr, and 100 mM DETA stock solutions were prepared. TTABr was purchased from Sigma-Aldrich (St. Louis, MO) and CTABr, DETA, and HDMBr were obtained from Aldrich Chemical Company Inc. (Milwaukee, WI). The EOF modifiers were converted to their hydroxide form by passing the solution through an anion exchange cartridge (Maxi-clean ICOH plus, Alltech Associates). The separation buffers were prepared by mixing the electrolyte solutions, EOF modifier solutions and DI water to obtain the desired concentrations. The buffer solutions were pH adjusted, filtered, and degassed before use.

Table 3.3
Phosphate buffer-HDMOH experimental matrix

	12.5 mM phosphate (6.25 mM each of Na_2HPO_4 and NaH_2PO_4)			
HDMOH	pH 7	pH 5.7	pH 4	pH 3.4
5 mg/L	×		×	
10 mg/L	×	×	×	×
20 mg/L	×	×	×	×
	25 mM phosphate (12.5 mM each of Na_2HPO_4 and NaH_2PO_4)			
5 mg/L	×	×	×	×

RESULTS

Phosphate Buffer

Equimolar concentrations of NaH_2PO_4 and Na_2HPO_4 were used to prepare the phosphate buffer. Four EOF modifiers were tested: HDMOH, CTAOH, DETA, and TTAOH. Pressure injection of 200 mbar•s was used to load samples onto the capillary.

Phosphate and HDMOH

The pH and the concentrations of phosphate and HDMOH were varied. The concentration and pH ranges selected were: 12.5 mM and 25 mM for phosphate, 5 to 20 mg/L for HDMOH, and pH 3.4 to 7 (Table 3.3).

Changing the buffer pH affected the peak resolution and peak shape significantly while changing the HDMOH concentration and the phosphate concentration had little effect on those parameters. The best results, with tall and sharp peaks, were obtained for the following conditions: 25 mM phosphate, 5 mg/L HDMOH, pH 5.7 (Figure 3.1). Nevertheless, dibromoacetic acid (DBAA) and trichloroacetic acid (TCAA) comigrated and only nine peaks were obtained.

At pH 7, there was a peak broadening problem for the chloroacetic acid (CAA), bromoacetic acid (BAA), and dicholoacetic acid (DCAA) peaks (results not shown). At the lower pH values of 3.4 and 4, there were more comigration problems than at pH 5.7 as only seven peaks were obtained (results not shown).

Phosphate and DETA

The pH and concentration of phosphate electrolyte (12.5 mM and 25 mM) were varied. The pH values tested using a phosphate concentration of 12.5 mM included: 4.00,4.96, 5.30, 5.50, 5.70, 5.80, 6.02, 7.00, and 9.25. The pH values tested using the 25 mM phosphate solution included: 5.70, 7.00, and 7.96. Electrolytes were freshly prepared, filtered (0.2 μm nylon membrane filter), and vacuum degassed. The pH of the solution was adjusted with phosphoric acid.

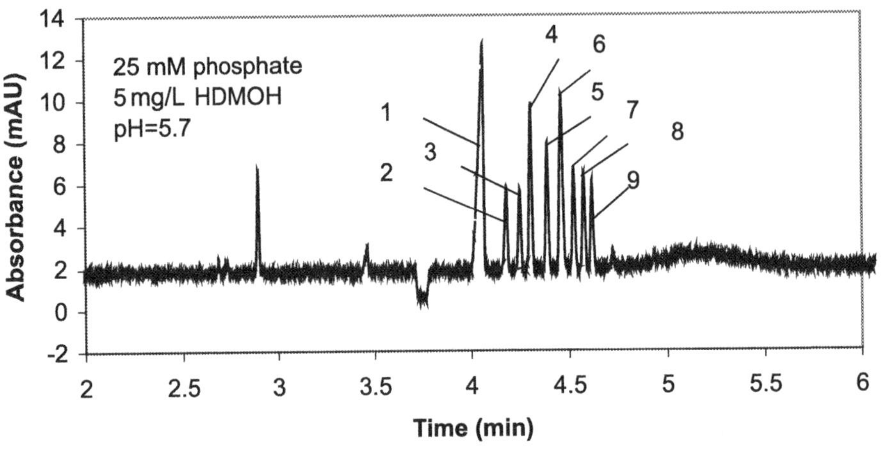

Figure 3.1 CE electropherogram from the analysis of a composite HAA solution using the phosphate-HDMOH buffer system
(Sample composition: 10 ppm HAAs and 100 ppm TFAA; Peak identification: 1-TFAA, 2-CAA, 3-BAA, 4-DCAA, 5-BCAA, 6-DBAA and TCAA, 7-BDCAA, 8-CDBAA, 9-TBAA; Buffer composition: 25 mM phosphate, 5 mg/L HDMOH, pH 5.7; Run conditions: Temperature = 25°C, Voltage = –20 kV, Injection = pressure (200 mbar·s), Detection = direct UV at 191 nm; Capillary: straight capillary (75 μm ID × 56 cm effective length))

Changing buffer pH affected the peak resolution and peak shape significantly while changing phosphate concentration had little effect. There was co-migration of TCAA and DBAA at all pH values. The optimum results in terms of peak resolution were obtained for the following conditions: 12.5 mM phosphate, 5 mM DETA, pH 5.7. At pH 5.7, all ten sample peaks were observed; however, TCAA and DBAA could not be completely separated (Figure 3.2). At pH 5.5, the peak migration times were greater than at pH 5.7 (i.e., longer run times) and the TCAA and DBAA peaks comigrated (Figure 3.2). However, the baseline at the lower pH was more stable and sharp peaks were obtained.

Phosphate and CTAOH

The concentrations of phosphate and CTAOH were set to 12.5 mM and 1.0 mM, respectively. The pH of the electrolyte was adjusted to the test values (4, 5.7, and 7) with phosphoric acid. Poor HAA separation was observed for phosphate in combination with CTAOH. For example, only 5 peaks were obtained for the following conditions: 12.5 mM phosphate, 1.0 mM CTAOH, pH=5.7 (Figure 3.3). In addition, the run times were longer than for the other EOF modifiers and the baseline was not as stable.

Phosphate and TTAOH

Equimolar concentrations of NaH_2PO_4 and Na_2HPO_4 (12.5 mM each) were dissolved in distilled and deionized (DI) water to prepare the phosphate buffer. TTAOH was prepared from the brominated form (TTABr) by passing a solution of 1.0 mM TTABr through an Alltech Maxi-clean IC-OH plus ion exchange cartridge. The pHs tested were: 4, 5.7, and 7.

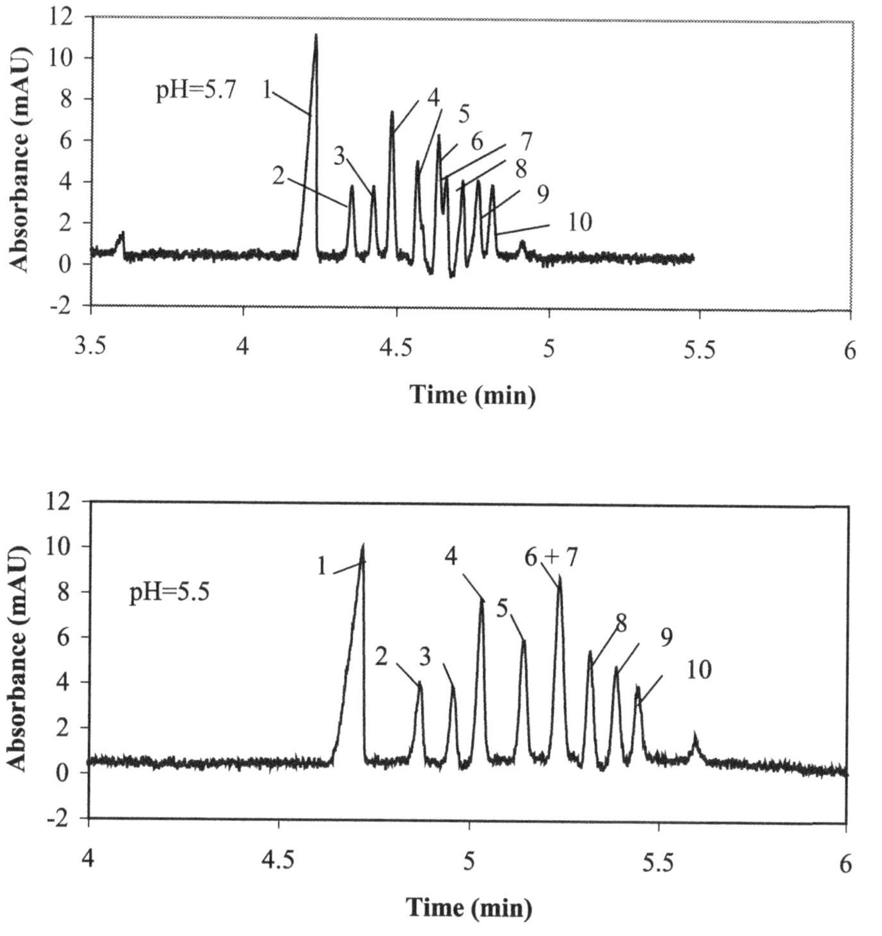

Figure 3.2 Effect of pH on separation of HAAs using a phosphate-DETA buffer system (Sample composition: 10 ppm HAAs and 100 ppm TFAA; Peak identification: 1-TFAA, 2-CAA, 3-BAA, 4-DCAA, 5-BCAA, 6-DBAA, 7-TCAA, 8-BDCAA, 9-CDBAA, 10-TBAA; Buffer composition: 12.5 mM phosphate, 5 mM DETA, pH 5.7; Run conditions: Temperature = 25°C, Voltage = –20 kV, Injection = pressure (200 mbar·s), Detection = direct UV at 191 nm; Capillary: straight capillary (75 μm ID × 56 cm effective length))

The optimum results were obtained for the following conditions: 25 mM phosphate, 1 mM TTAOH, pH 5.7 (Figure 3.4). Complete separation of the HAAs was observed and sharp peaks were obtained for all compounds.

Citric Acid and LiOH Buffer

The concentration of citric acid was fixed at 50 mM and the concentration of LiOH was varied from 10 to 70 mM. HDMOH (5 mg/L) served as the EOF modifier. The solution pH decreased as the concentration of LiOH was increased (Table 3.4). Pressure injection of 200 mbar·s was used to load samples onto the capillary.

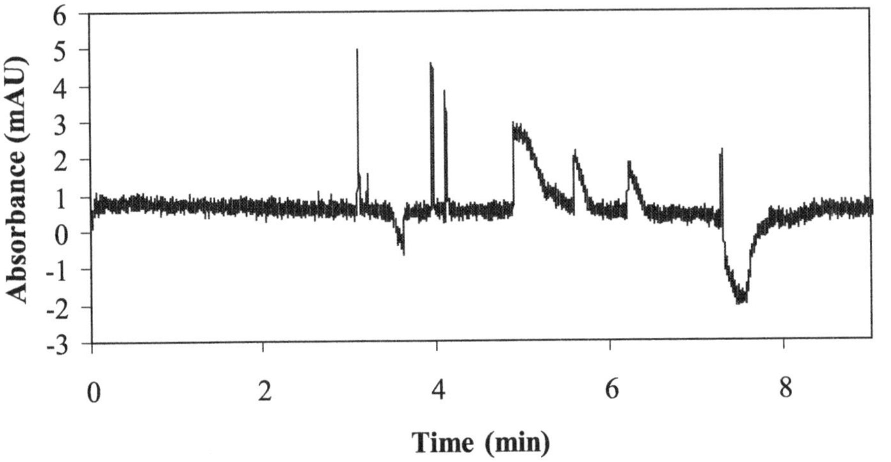

Figure 3.3 Separation of HAAs using a phosphate-CTAOH buffer system
(Sample composition: 10 ppm HAAs and 100 ppm TFAA; Buffer composition: 12.5 mM phosphate, 1 mM CTAOH, pH = 5.7; Run conditions: Temperature = 25°C, Voltage = –20 kV, Injection = pressure (200 mbar·s), Detection = direct UV at 191 nm; Capillary: straight capillary (75 μm ID × 56 cm effective length))

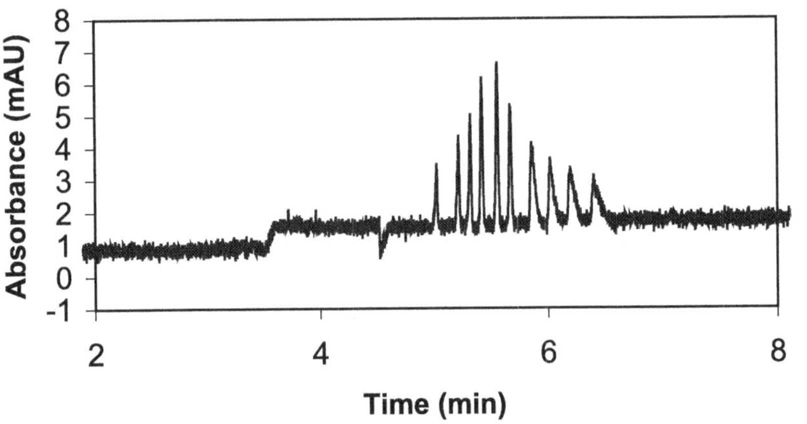

Figure 3.4 Separation of HAAs using a phosphate-TTAOH buffer system
(Sample composition: 10 ppm HAAs and 10 ppm TFAA; Peak identification: 1-TFAA, 2-CAA, 3-BAA, 4-DCAA, 5-BCAA, 6-DBAA, 7-TCAA, 8-BDCAA, 9-CDBAA, 10-TBAA; Buffer composition: 12.5 mM NaH_2PO_4, 12.5 mM Na_2HPO_4, 1 mM TTAOH, pH = 5.7; Run conditions: Temperature = 25°C, Voltage = –20 kV, Injection = pressure (200 mbar·s), Detection = direct UV at 191 nm; Capillary: straight capillary (75 μm ID × 56 cm effective length))

Table 3.4
Experimental matrix for testing the citric acid + LiOH with HDMOH buffer system

Run number	Citric acid (mM)	LiOH (mM)	HDMOH (mg/L)	pH
1	50	10	10	2.77
2	50	25	10	3.15
3	50	35	10	3.40
4	50	50	10	3.81
5	50	70	10	4.37

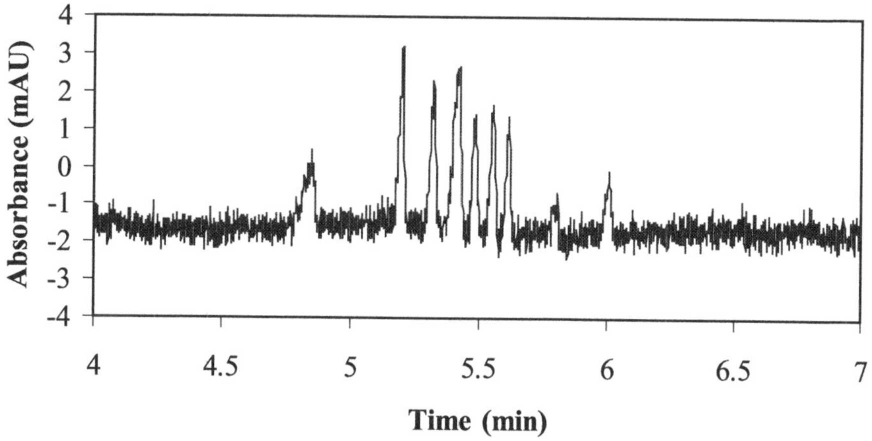

Figure 3.5 Separation of HAAs using a citric acid+LiOH-HDMOH buffer system (Sample composition: 10 ppm HAAs and 100 ppm TFAA; Buffer composition: 50 mM citric acid, 35 mM LiOH, 10 mg/L HDMOH, pH = 3.4; Run conditions: Temperature = 25°C, Voltage = –20 kV, Injection = pressure (200 mbar·s), Detection = direct UV at 191 nm; Capillary: straight capillary (75 µm ID × 56 cm effective length))

Poor separation of HAAs was observed using the citric acid + LiOH with HDMOH buffer system. For example, the electropherogram for run 3 (pH=3.4) only yielded six sharp peaks (Figure 3.5).

NDC Buffer

NDC is the only buffer in the experimental plan (Table 3.1) that functions by indirect detection. NDC provides a high baseline absorbance (especially around 235 nm) and the sample peaks are "negative" or inverted for compounds that have a negligible absorbance at the detection wavelength.

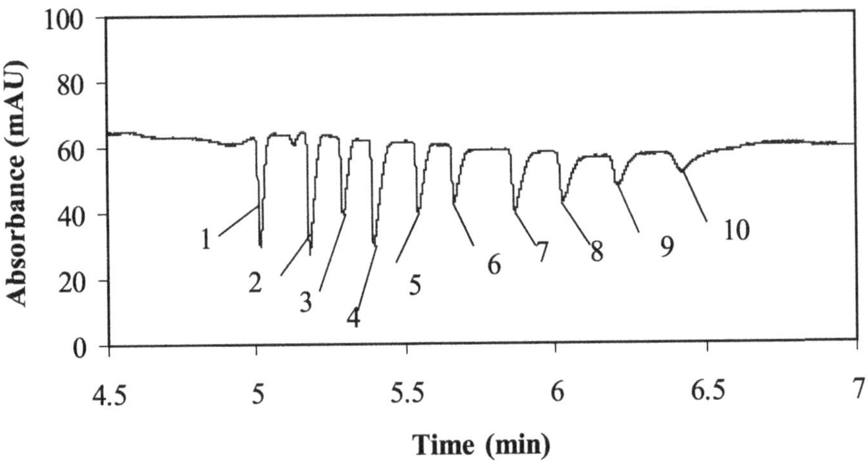

Figure 3.6 Separation of HAAs using an NDC-CTABr buffer system
(Peak identification: 1-TFAA, 2-CAA, 3-BAA, 4-DCAA, 5-BCAA, 6-DBAA, 7-TCAA, 8-BDCAA, 9-CDBAA, 10-TBAA; Buffer composition: 4 mM NDC, 0.5 mM CTABr, pH = 7.5; Run conditions: Temperature = 25°C, Voltage = –20 kV, Injection = pressure (200 mbar•s), Detection = indirect UV at 235 nm; Capillary: straight capillary (75 μm ID × 56 cm effective length))

Effect of Different EOF Modifiers

Four EOF modifiers were tested: 0.5 mM CTABr, 5 mM DETA, 5 mg/L HDMBr, and 1 mM TTABr. The concentration of NDC was 4 mM and the pH of the buffer was 7.5. The concentrations of nine HAAs and TFAA were 10 mg/L and 10 mg/L, respectively. The detector wavelength was set at 235 nm for indirect detection. Pressure injection of 200 mbar•s was used to load samples onto the capillary.

Complete separation of HAAs and TFAA was observed with CTABr as the EOF modifier (Figure 3.6). There was peak broadening for BDCAA, CDBAA and TBAA. No peaks were obtained when DETA was used as the EOF modifier (Figure 3.7). A loss in peak resolution was observed when HDMBr or TTABr served as the EOF modifier (Figures 3.8 and 3.9). The results suggest that NDC and CTABr are a suitable combination for separating the nine HAAs and TFAA. Martinez et al. (1999) also observed that NDC and CTABr buffer combination provided the optimum HAA peak resolution when sodium chromate, potassium hydrogenphthalate, NDC, and sodium tetraborate buffer systems were compared.

Based on the results discussed above, it can be concluded that two buffer systems provided acceptable peak resolution and peak shape for HAAs. One buffer system was comprised of 12.5 mM NaH_2PO_4, 12.5 mM Na_2HPO_4, and 1 mM TTAOH at pH 5.7. The other buffer system was comprised of 4 mM NDC and 1 mM CTABr at pH 7.5. Thus, these two buffer systems were selected for future evaluation and method optimization.

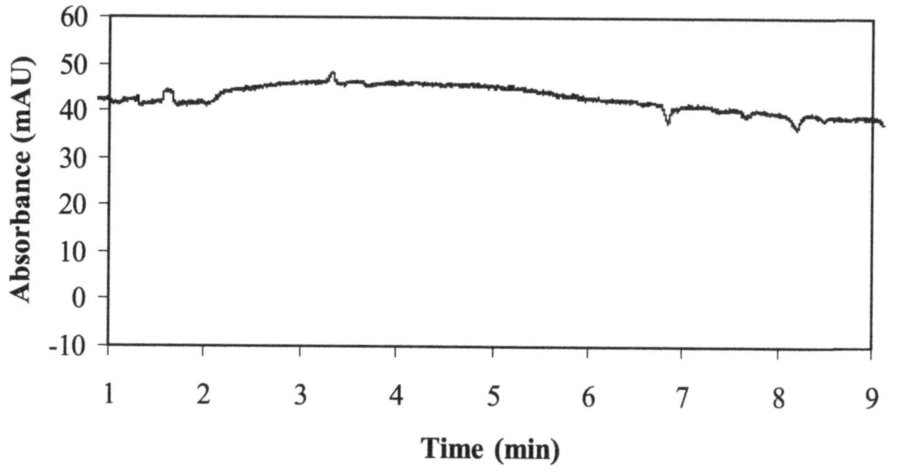

Figure 3.7 Separation of HAAs using an NDC-DETA buffer system
(Buffer composition: 4 mM NDC, 5 mM DETA, pH = 7.5; Run conditions: Temperature = 25°C, Voltage = –20 kV, Injection = pressure (200 mbar·s), Detection = indirect UV at 235 nm; Capillary: straight capillary (75 μm ID × 56 cm effective length))

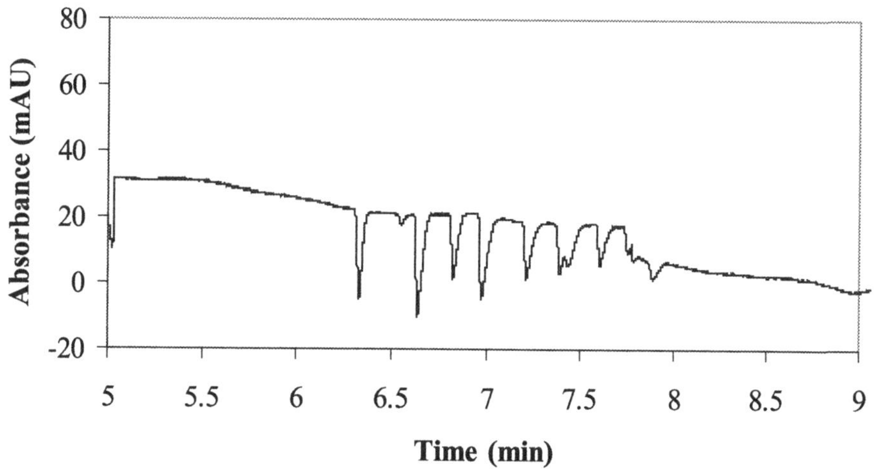

Figure 3.8 Separation of HAAs using an NDC-HDMBr buffer system
(Buffer composition: 4 mM NDC, 5 mg/L HDMBr, pH = 7.5; Run conditions: Temperature = 25°C, Voltage = –20 kV, Injection = pressure (200 mbar·s), Detection = indirect UV at 235 nm; Capillary: straight capillary (75 μm ID × 56 cm effective length))

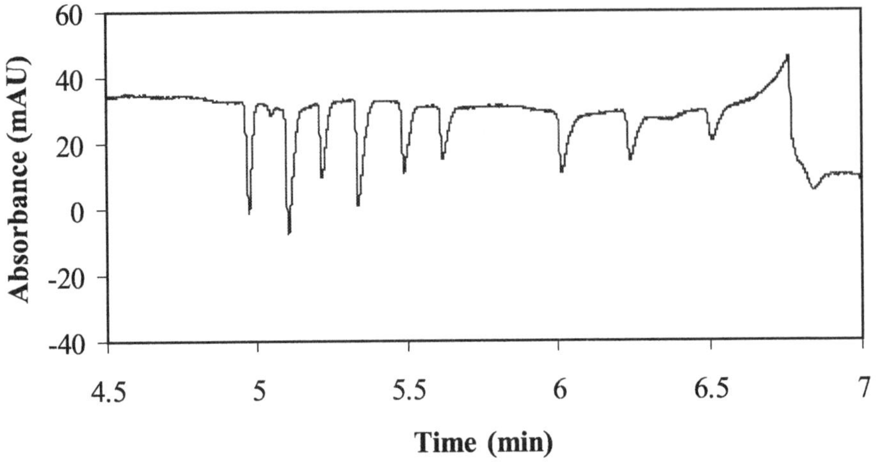

Figure 3.9 Separation of HAAs using an NDC-TTABr buffer system
(Buffer composition: 4 mM NDC, 1 mM TTABr, pH = 7.5; Run conditions: Temperature = 25°C, Voltage = –20 kV, Injection = pressure (200 mbar·s), Detection = indirect UV at 235 nm; Capillary: straight capillary (75 μm ID × 56 cm effective length))

SUMMARY

Complete peak resolution and good peak shapes were obtained for all nine HAAs and TFAA using two buffer systems: 1) 4 mM NDC and 0.5 mM CTABr at pH 7.5 and 2) 12.5 mM NaH_2PO_4, 12.5 mM Na_2HPO_4 and 1 mM TTAOH at pH 5.7. A stable baseline was obtained using phosphate buffer; however, NDC buffer provided a relatively less stable baseline. Therefore, the remaining method development work primarily focused on the phosphate buffer system with some additional evaluation of the NDC buffer system.

CHAPTER 4
CALIBRATION CURVES AND IDLs FOR HAAs USING PHOSPHATE BUFFER WITH PRESSURE INJECTION

PREPARATION OF AQUEOUS STANDARD SOLUTIONS

Introduction

Aqueous standard solutions were used to determine the calibration curves and IDLs for HAAs. One difficulty with preparing standard solutions from neat materials is that several of the HAAs are extremely hygroscopic solids at room temperature. Therefore, it is difficult to accurately weigh out quantitative standards. The procedure used to prepare the aqueous standards is described below.

Materials and Methods

Neat dichloroacetic acid (DCAA) and trifluoroacetic acid (TFAA) (Sigma-Aldrich, St. Louis, MO) are liquids at room temperature. A 1000 mg/L DCAA aqueous stock solution was prepared by simply pipetting the desired volume of DCAA into a volumetric flask and diluting with distilled and deionized (DI) water. A 1000 mg/L TFAA aqueous standard was prepared following the same procedure.

Neat tribromoacetic acid (TBAA) (Sigma-Aldrich, St. Louis, MO) and chlorodibromoacetic acid (CDBAA) (Supelco, Bellefonte, PA) are not hygroscopic and are relatively dry solids. Therefore, TBAA was simply weighted out into a volumetric flask and dissolved with DI water to prepare a 1000 mg/L stock solution. The 1000 mg/L CDBAA aqueous stock solution was prepared in the same manner.

The remaining neat HAAs are hygroscopic solids and they appeared wet and clumped in the bottles shipped from the vendors (Sigma-Aldrich or Supelco). It is difficult to weigh out these compounds to make accurate aqueous standard solutions. Therefore, those chemicals were freeze dried (i.e., frozen in the presence of a vacuum) using a commercial freeze drying apparatus for 24 hours. Application of freeze drying to chloroacetic acid (CAA), bromoacetic acid (BAA), and dibromoacetic acid (DBAA) resulted in dry solids. Then, 1000 mg/L aqueous stock solutions of these compounds were prepared by simply weighing out the desired amount of dried neat chemical and dissolving with DI water in a volumetric flask.

Trichloroacetic acid (TCAA), bromochloroacetic acid (BCAA) and bromodichloroacetic acid (BDCAA) disappeared from the covered bottles during the freeze-drying operation. This likely occurred because these three HAAs are the most hygroscopic and they evaporated with the moisture during the drying period. It is also possible that they volatilized or sublimed upon application of a vacuum. Therefore, approximate amounts of TCAA, BCAA or BDCAA were weighed out and made to volume with DI water in a volumetric flask. The concentrations of these solutions were verified by a sodium hydroxide (NaOH) titration method. The detailed procedure is as follows. An aliquot (20 mL or 10 mL) of these aqueous solutions was transferred to a 50 mL beaker and titrated with diluted (0.0050 M) NaOH from a certified 0.1000 M NaOH solution (Fisher Scientific, New Jersey). Nitrogen was sparged into the acid solution during titration to avoid interference from carbon dioxide. The 0.0050 M NaOH solution was added to the HAA solution in increments of 1 mL or less (0.05 mL when close to the equivalence point). After each incremental addition, the pH was recorded when the reading stabilized. The titration was continued until the pH value exceeded 9.0. The titration curve of pH versus volume of titrant added was plotted and the inflection point was determined. Finally, the concentrations of the TCAA, BDCAA, and BCAA aqueous stock solutions were computed as follows:

Concentration (mg/L) = [vol. titrant (mL) × 5 mM × HAA MW (mg/mmole)]/vol. sample (mL)

Results

Titration curves for TCAA, BDCAA and BCAA are shown in Figures 4.1, 4.2, and 4.3, respectively. The inflection points occurred at pH 7, which indicates that TCAA, BDCAA and BCAA are relatively strong acids. The concentrations of the aqueous stock of TCAA, BDCAA and BCAA solutions are shown in Table 4.1. The observed concentrations were similar to the 1000 mg/L target concentration suggesting that moisture issues were not very important. Fortunately, the titration method is reproducible and appears to be a viable method for verifying the concentrations of HAA stock solutions.

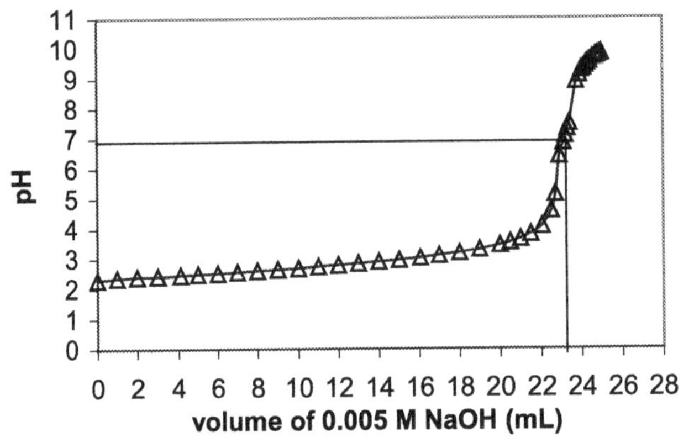

Figure 4.1 Titration curve for 20 mL of TCAA (0.0050 M NaOH as titrant)

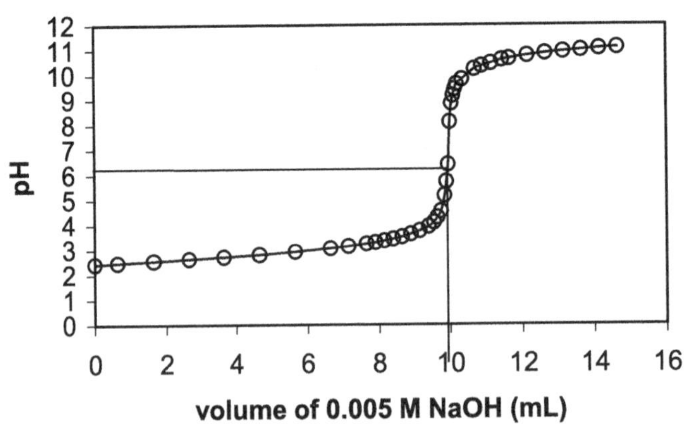

Figure 4.2 Titration curve for 10 mL of BDCAA (0.0050 M NaOH as titrant)

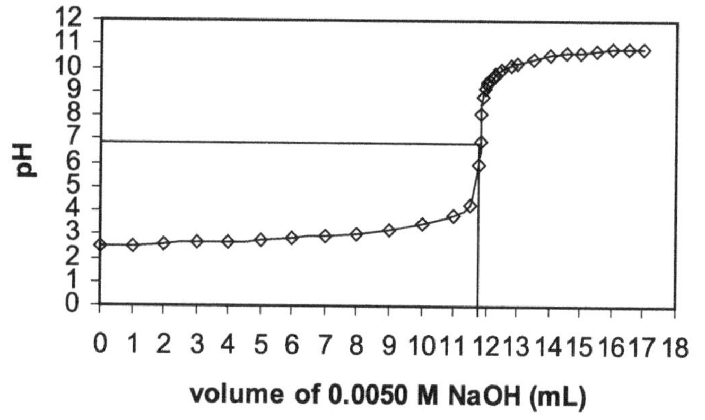

Figure 4.3 Titration curve for 10 mL of BCAA (0.0050 M NaOH as titrant)

Table 4.1
Concentrations of aqueous stock solutions of TCAA, BDCAA and BCAA

Acid name	Volume of acid stock solution for titration (mL)	Volume of 0.0050 M NaOH used (mL)		Concentration of acid stock solution ($\times 10^{-3}$ M)		Average conc. (mg/L)	Relative standard deviation* (%)
		Titration A	Titration B	A	B		
TCAA	20	23.15	23	5.788	5.75	942.6	0.45
BDCAA	10	9.975	-	4.988	-	1037	-
BCAA	10	11.8	11.5	5.9	5.75	1011	1.82

* Relative standard deviation (RSD) is calculated as 100 × standard deviation of two measurements/average of two measurements.

DETERMINATION OF THE CALIBRATION CURVES AND IDLs FOR HAAs

Straight Capillary

Separate calibration standards containing 2 mg/L, 5 mg/L, 10 mg/L, 25 mg/L, and 40 mg/L of one HAA in DI water were prepared from the individual 1000 mg/L aqueous stock solutions. The buffer system consisted of the following: 12.5 mM Na_2HPO_4, 12.5 mM NaH_2PO_4, 1 mM TTAOH, pH = 5.7. A straight capillary (75 μm ID × 56 cm effective length) was used. The sample was applied onto the capillary with a pressure injection of 50.0 mbar for 4 s and the HAAs were detected at 191 nm. A representative calibration curve obtained for CAA is included as Figure 4.4. The calibration curves were linear over the range investigated. To determine the IDL, each HAA aqueous stock solution was diluted with DI water to a concentration near the estimated IDL. Seven replicate analyses of this solution were performed and the IDL was calculated by multiplying the standard deviation by 3.14 (APHA, AWWA, WEF 1998). The calibration equations for the HAAs and TFAA and the IDLs obtained using a straight capillary are summarized in Table 4.2.

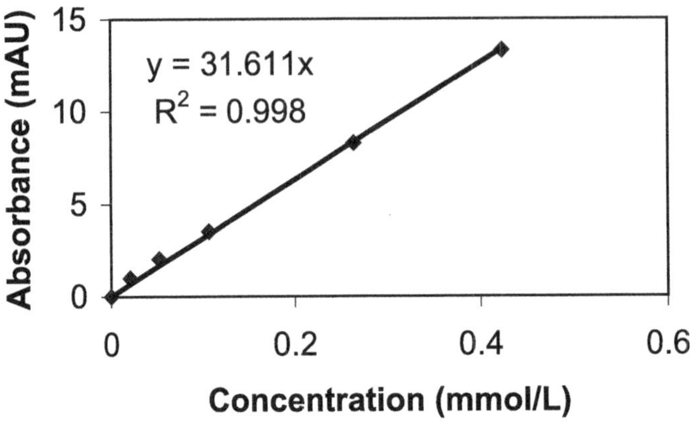

Figure 4.4 Calibration curve for CAA obtained using the straight capillary

Table 4.2
Calibrations curves and IDLs for each HAA using phosphate-TTAOH buffer and a straight capillary

Compound	Calibration equation	R^2	IDL (μM)	IDL (mg/L)
Trifluoroacetic acid	y=24.869x	0.9998	18.7	2.14
Chloroacetic acid	y=31.611x	0.998	20.8	1.96
Bromoacetic acid	y=55.51x	0.9962	12.7	1.76
Dichloroacetic acid	y=75.265x	0.9993	11	1.42
Bromochloroacetic acid	y=118.66x	0.9982	7.6	1.31
Dibromoacetic acid	y=119.7x	0.9984	7.9	1.73
Trichloroacetic acid	y=97.624x	0.9941	9.9	1.62
Bromodichloroacetic acid	y=108.58x	0.9983	7.9	1.65
Chlorodibromoacetic acid	y=153.66x	0.9996	7.2	1.82
Tribromoacetic acid	y=214.6x	0.9996	6.3	1.86

y = peak area (mAU•min)
x = HAA concentration (mM)

Bubble Cell Capillary

The bubble cell has a longer detection path length than the straight capillary; therefore, an increase in sensitivity is expected using the bubble cell design. Nevertheless, some dispersion is expected in the bubble cell, which could lead to a loss in peak resolution. Therefore, the separation of the nine HAAs and TFAA was evaluated using a bubble cell capillary. Complete separation of the nine HAAs and TFAA was observed using a bubble cell capillary (Figure 4.5)

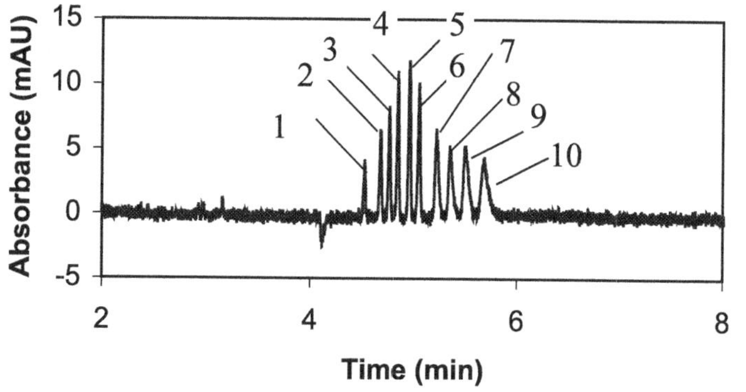

Figure 4.5 Separation of 10 mg/L HAAs and 10 mg/L TFAA using a bubble cell capillary (Peak identification: 1-TFAA, 2-CAA, 3-BAA, 4-DCAA, 5-BCAA, 6-DBAA, 7-TCAA, 8-BDCAA, 9-CDBAA, 10-TBAA; Buffer composition: 12.5 mM NaH_2PO_4, 12.5 mM Na_2HPO_4, 1 mM TTAOH, pH = 5.7; Run conditions: Temperature = 25°C, Voltage = –20 kV, Injection = pressure (200 mbar•s), Detection = direct UV at 191 nm; Capillary: bubble cell (75 µm ID × 56 cm effective length))

and there was an increase in peak height compared to the straight capillary (compare Figures 4.5 and Figure 3.4). The buffer system consisted of the following: 12.5 mM Na_2HPO_4, 12.5 mM NaH_2PO_4, 1 mM TTAOH, pH = 5.7. Pressure injection of 50.0 mbar for 4 s was used.

The same calibration standards, buffer conditions, injection method and detection wavelength were used to obtain calibration curves for the HAAs using a bubble cell capillary. The calibration equations and IDLs for the nine HAAs and TFAA using phosphate and TTAOH buffer with a bubble cell capillary are summarized in Table 4.3. Pressure injection of 50 mbar for 4 seconds was used to load samples onto the capillary.

Comparing the calibration curve slopes for each HAA using the straight capillary and the bubble cell capillary (Table 4.4), there is approximately a 3-fold increase in sensitivity when the bubble cell is used.

An injection is typically described by the product of the applied pressure and the time that the pressure was applied. A greater amount of sample is loaded onto the column if the sample injection time or pressure or both are increased, thereby increasing the method sensitivity. However, the separation efficiency may decrease if excessive sample fills a large portion of the capillary length and the capillary length available for analyte separation decreases. The effect of a 3-fold increase in injection time for a 600 mbar•s injection (50 mbar for 12 seconds) was evaluated. As shown in Figure 4.6, all nine HAAs and TFAA were still completely separated at pressure injection of 600 mbar•s. The calibration equations and IDLs for the nine HAAs and TFAA using phosphate and TTAOH buffer with a bubble cell capillary and pressure injection of 600 mbar•s are summarized in Table 4.5. Comparing the calibration curve slopes for each HAA using pressure injection of 200 mbar•s and pressure injection of 600 mbar•s (Table 4.6), there is approximately a 3-fold increase in sensitivity when pressure injection of 600 mbar•s was used to load the sample onto the capillary. Further increasing to 1000 mbar•s was not successful because comigration of some peaks observed for 40 mg/L HAAs (Figure 4.7).

Table 4.3
Calibration curves for each HAA using phosphate + TTAOH buffer and a bubble cell capillary

Compound	Calibration equation	R^2	IDL (μM)	IDL (mg/L)
Trifluoroacetic acid	y=73.34x	0.9966	6.85	0.78
Chloroacetic acid	y=87.561x	0.9941	7.62	0.72
Bromoacetic acid	y=149.72x	0.9975	4.65	0.64
Dichloroacetic acid	y=216.94x	0.9992	4.03	0.52
Bromochloroacetic acid	y=324.03x	0.9996	2.78	0.48
Dibromoacetic acid	y=363.65x	0.9992	2.89	0.63
Trichloroacetic acid	y=266.56x	0.999	3.63	0.59
Bromodichloroacetic acid	y=295.61x	0.9973	2.91	0.60
Chlorodibromoacetic acid	y=421.0x	0.9972	2.64	0.67
Tribromoacetic acid	y=631.76x	0.9983	2.31	0.68

y = peak area (mAU•min)
x = HAA concentration (mM)

Table 4.4
Comparison of calibration curve slopes obtained with the straight and bubble cell capillaries

Compound	Calibration curve slope		Ratio
	Straight	Bubble cell	
Trifluoroacetic acid	24.869	73.34	2.95
Chloroacetic acid	31.611	87.561	2.77
Bromoacetic acid	55.51	149.72	2.70
Dichloroacetic acid	75.265	216.94	3.00
Bromochloroacetic acid	118.66	324.03	2.73
Dibromoacetic acid	119.7	363.65	3.04
Trichloroacetic acid	97.624	266.56	2.73
Bromodichloroacetic acid	108.58	295.61	2.72
Chlorodibromoacetic acid	153.66	421.0	2.74
Tribromoacetic acid	214.6	631.76	2.94

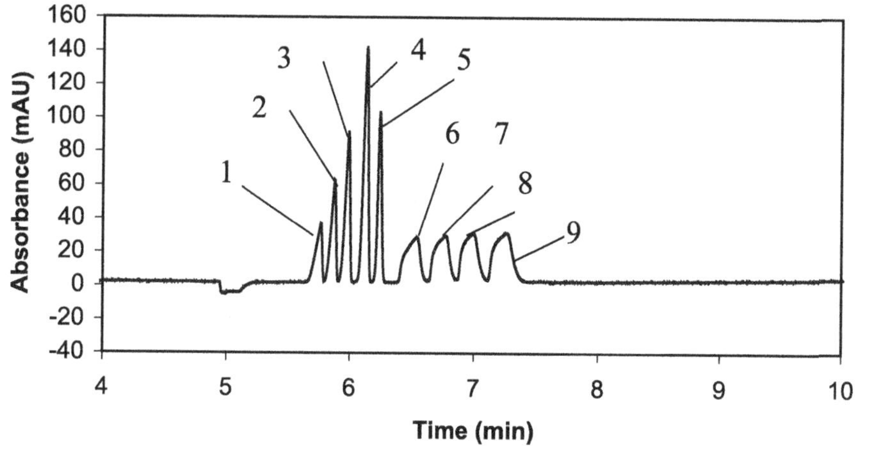

Figure 4.6 Separation of 40 mg/L HAAs using a bubble cell capillary (Peak identification: 1-CAA, 2-BAA, 3-DCAA, 4-BCAA, 5-DBAA, 6-TCAA, 7-BDCAA, 8-CDBAA, 9-TBAA; Buffer composition: 12.5 mM NaH_2PO_4, 12.5 mM Na_2HPO_4, 1 mM TTAOH, pH = 5.7; Run conditions: Temperature = 25°C, Voltage = –20 kV, Injection = pressure (600 mbar·s), Detection = direct UV at 191 nm; Capillary: bubble cell (75 µm ID × 56 cm effective length))

Table 4.5
Calibration curves for each HAA using phosphate + TTAOH buffer and a bubble cell capillary

Compound	Calibration equation	R^2	IDL (µM)	IDL (mg/L)
Trifluoroacetic acid	y=185.9x	0.9931	2.38	0.27
Chloroacetic acid	y=267.9x	0.998	2.65	0.25
Bromoacetic acid	y=491.5x	0.9983	1.62	0.22
Dichloroacetic acid	y=696.6x	0.9995	1.40	0.18
Bromochloroacetic acid	y=965.7x	0.9992	0.97	0.17
Dibromoacetic acid	y=1159.4x	0.9992	1.00	0.22
Trichloroacetic acid	y=886.5x	0.9994	1.26	0.20
Bromodichloroacetic acid	y=966.9x	0.9991	1.01	0.21
Chlorodibromoacetic acid	y=1281.4x	0.999	0.92	0.23
Tribromoacetic acid	y=1801.9x	0.999	0.80	0.24

y = peak area (mAU•min)
x = HAA concentration (mM)

Table 4.6
Comparison of calibration curve slopes obtained with the bubble cell capillaries using pressure injection at 200 mbar·s or 600 mbar·s

	Calibration curve slope		
Compound	200 mbar·s	600 mbar·s	Ratio
Trifluoroacetic acid	67	185.9	2.77
Chloroacetic acid	105	267.9	2.55
Bromoacetic acid	169	491.5	2.91
Dichloroacetic acid	243	696.6	2.87
Bromochloroacetic acid	329	965.7	2.94
Dibromoacetic acid	398	1159.4	2.91
Trichloroacetic acid	304	886.5	2.92
Bromodichloroacetic acid	337	966.9	2.87
Chlorodibromoacetic acid	440	1281.4	2.91
Tribromoacetic acid	615	1801.9	2.93

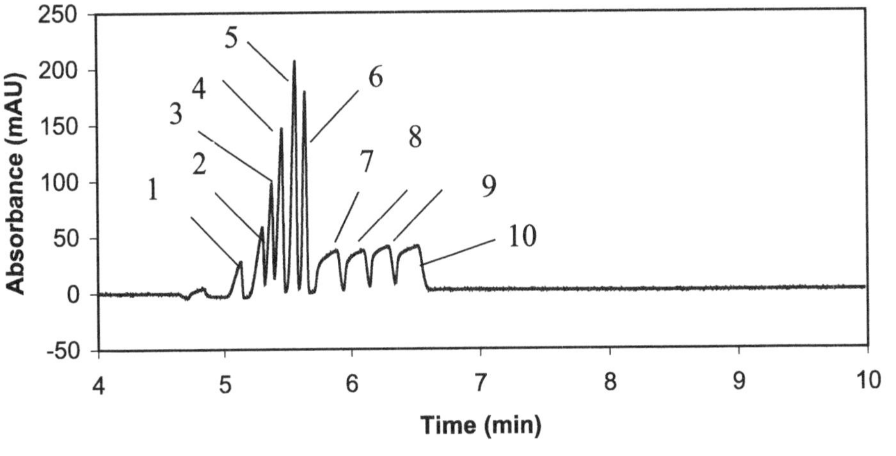

Figure 4.7 Separation of 40 mg/L HAAs using a bubble cell capillary
(Peak identification: 1-TFAA, 2-CAA, 3-BAA, 4-DCAA, 5-BCAA, 6-DBAA, 7-TCAA, 8-BDCAA, 9-CDBAA, 10-TBAA; Buffer composition: 12.5 mM NaH_2PO_4, 12.5 mM Na_2HPO_4, 1 mM TTAOH, pH = 5.7; Run conditions: Temperature = 25°C, Voltage = –20 kV, Injection = pressure (1000 mbar·s), Detection = direct UV at 191 nm; Capillary: bubble cell (75 μm ID × 56 cm effective length))

SUMMARY

Calibration curves for all nine HAAs and TFAA were linear in the range of 0 to 40 mg/L using phosphate+TTAOH buffer and a straight capillary or a bubble cell capillary with pressure injection. There was approximately a 3-fold increase in sensitivity when the bubble cell was used compared to a straight capillary. Complete separation of all nine HAAs and TFAA was obtained when a bubble cell capillary with pressure injection of 600 mbar•s was used and a 3-fold increase in sensitivity was obtained in comparison to a pressure injection of 200 mbar•s. Thus, phosphate +TTAOH buffer and a bubble cell capillary with pressure injection of 600 mbar•s was selected for further CE method. Because the IDLs for the HAAs were in the range of 0.17 to 0.27 mg/L using this method, more work is needed to achieve the low μg/L MDLs required for a viable method.

CHAPTER 5
SAMPLE CONCENTRATION USING ON-CAPILLARY CONCENTRATION TECHNIQUES AND OFF-LINE SOLID-PHASE EXTRACTION

INTRODUCTION

The IDLs for HAAs were in the range of 170 to 270 µg/L using phosphate + TTAOH buffer, a bubble cell capillary and pressure injection of 600 mbar•s. These IDLs were much greater than the MDLs for the EPA-approved methods (1 µg/L). In order to reach the method detection limit (MDL) goal of 1 µg/L, on-capillary concentration or off-line sample concentration was needed. Electrokinetic injection and CITP are two common on-capillary concentration techniques used in CE. Off-line concentration options include liquid-liquid extraction and solid-phase extraction. Liquid-liquid extraction was not investigated in this project as it requires the use of potentially harmful organic solvents, is time-consuming, and produces an extract which may not be suitable for injection into a CE ion analyzer. Thus, electrokinetic injection and CITP as the on-capillary concentration approaches and solid-phase extraction as the off-line approach were evaluated in this phase of the project.

EVALUATION OF ELECTROKINETIC INJECTION

All previous analyses were performed using the most common injection method, pressure injection (200 mbar•s). For a pressure injection, pressure is applied to the sample reservoir placed at the inlet end of the capillary to force the sample into the capillary. For a given applied pressure and time, the volume of sample loaded onto the capillary can be easily calculated using the Hagen-Poisseuille equation. Thus, pressure injection is comparable to using a syringe to apply a known volume of sample to an analytical instrument. On the other hand, electrokinetic injection involves application of a voltage to the system to introduce the analytes into the capillary via electrostatic attraction. As we discussed in Chapter 1, cations or anions can be preferentially introduced into the system when electrokinetic injection is used. This typically leads to a magnification in the observed signal (i.e., improved sensitivity) for the ions of interest. The effect of electrokinetic injection on method sensitivity was evaluated with 4 mM NDC and 0.5 mM CTABr as the buffer system and 12.5 mM NaH_2PO_4, 12.5 mM Na_2HPO_4 and 1 mM TTAOH as the buffer system. A pressure injection of 200 mbar•s and an electrokinetic injection of –5 kV for 20 seconds (–100 kV•s) were compared.

Phosphate Buffer

When electrokinetic injection (–5 kV for 20 s) was used to introduce the sample onto the capillary, complete separation of the nine HAAs and TFAA was also observed (Figure 5.1). For this analysis, a 1 mg/L HAA and TFAA composite standard was prepared from aqueous stock solutions. Sharp peaks were obtained for the mono- and dihalogenated species while peak broadening and peak distortion were observed for the trihalogenated species using electrokinetic

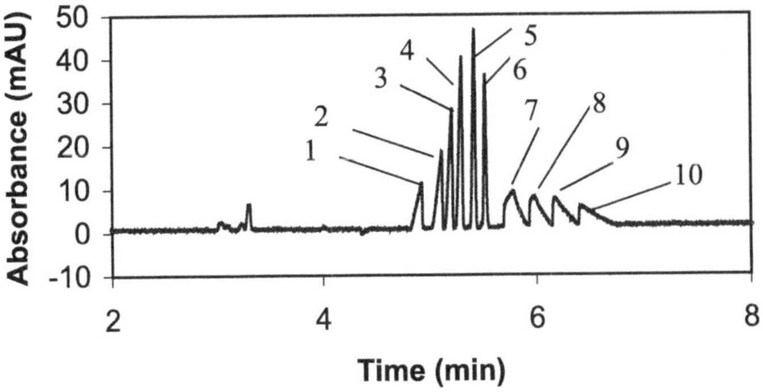

Figure 5.1 Separation of 1 mg/L HAAs and 1 mg/L TFAA following electrokinetic injection (Peak identification: 1-TFAA, 2-CAA, 3-BAA, 4-DCAA, 5-BCAA, 6-DBAA, 7-TCAA, 8-BDCAA, 9-CDBAA, 10-TBAA; Buffer composition: 12.5 mM NaH_2PO_4, 12.5 mM Na_2HPO_4, 1 mM TTAOH, pH = 5.7; Run conditions: Temperature = 25°C, Voltage = –20 kV, Injection = electrokinetic (–5 kV for 20 s), Detection = direct UV at 191 nm; Capillary: straight capillary (75 μm ID × 56 cm effective length))

injection (Figure 5.1). There was about a 1 to 2 order of magnitude increase in sensitivity for all compounds using electrokinetic injection compared to pressure injection.

NDC Buffer

As shown in Figure 5.2, complete separation of nine HAAs and TFAA was obtained for NDC buffer with electrokinetic injection. In addition, significant improvement in sensitivity was also obtained using electrokinetic injection (Figure 5.2) compared to pressure injection (Figure 5.3). The concentrations of each of the nine HAAs in the composite sample (dissolved in DI water) was 2 mg/L and the concentration of TFAA was 4 mg/L.

The effect of a gradient voltage on method sensitivity was also evaluated for the NDC buffer system. For the gradient voltage run, the voltage was initially set at –20 kV for 4.5 minutes and then decreased linearly to –15 kV over 0.5 minutes. Martinez et al. (1999) used this gradient voltage approach for the analysis of HAAs in swimming pool water using NDC buffer. Electrokinetic injection provided an approximately 20-fold increase in sensitivity over that obtained with pressure injection. There was an additional improvement in sensitivity for the di- and trihalogenated HAAs when the gradient voltage was used in conjunction with electrokinetic injection (Figure 5.4).

Electrokinetic injection yielded a significant improvement in sensitivity. However, one concern with electrokinetic injection is the potential for non-linear calibration curves and sample matrix effects. Therefore, these potential problems were evaluated using both NDC buffer and phosphate buffer.

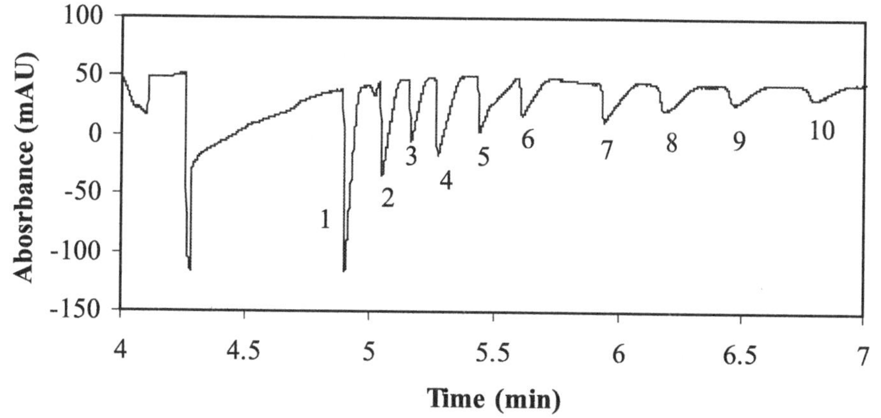

Figure 5.2 Separation of 2 mg/L HAAs and 4 mg/L TFAA following electrokinetic injection (Peak identification: 1-TFAA, 2-CAA, 3-BAA, 4-DCAA, 5-BCAA, 6-DBAA, 7-TCAA, 8-BDCAA, 9-CDBAA, 10-TBAA; Buffer composition: 4 mM NDC, 0.5 mM CTABr, pH = 7.5; Run conditions: Temperature = 25°C, Voltage = –20 kV, Injection = electrokinetic (–5 kV for 20 seconds), Detection = indirect UV at 235 nm; Capillary: straight capillary (75 μm ID × 56 cm effective length))

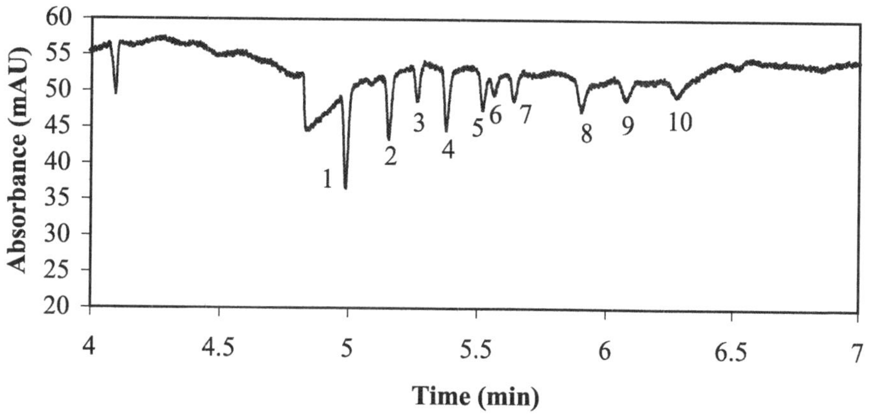

Figure 5.3 Separation of 2 mg/L HAAs and 4 mg/L TFAA following pressure injection (Peak identification: 1-TFAA, 2-CAA, 3-BAA, 4-DCAA, 5-BCAA, 6-DBAA, 7-TCAA, 8-BDCAA, 9-CDBAA, 10-TBAA; Buffer composition: 4 mM NDC, 0.5 mM CTABr, pH = 7.5; Run conditions: Temperature = 25°C, Voltage = –20 kV, Injection = pressure (200 mbar·s), Detection = indirect UV at 235 nm; Capillary: straight capillary (75 μm ID × 56 cm effective length))

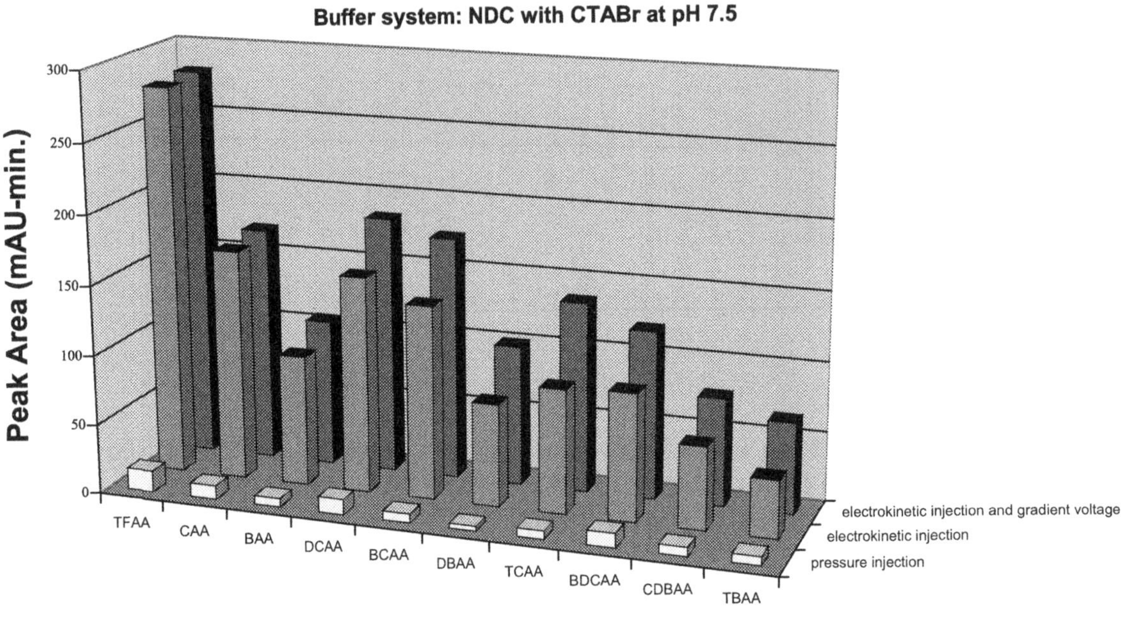

Figure 5.4 Effect of injection mode and the use of gradient voltage on HAA peak area (Sample: 2 mg/L HAAs and 4 mg/L TFAA); (Buffer composition: 4 mM NDC, 0.5 mM CTABr, pH = 7.5; Run conditions: Temperature = 25°C, Voltage = –20 kV, Injection = pressure (200 mbar·s), Detection = indirect UV at 235 nm; Capillary: straight capillary (75 µm ID 56 cm effective length))

DETERMINATION OF HAA CALIBRATION CURVES USING ELECTROKINETIC INJECTION

Phosphate Buffer

A composite standard containing 1 mg/L of each HAA and TFAA was prepared from individual aqueous 1000 mg/L HAA standard solutions. Then, separate composite standards containing 50, 100, 200, 250, and 500 µg/L of each HAA and TFAA were prepared from the 1 mg/L standard solution. Those standards were subsequently analyzed using capillary electrophoresis. The buffer contained 12.5 mM NaH_2PO_4, 12.5 mM Na_2HPO_4, and 1 mM TTAOH at pH=5.7. A straight capillary (75 µm ID × 56 cm effective length) was used. Electrokinetic injection (–5 kV for 20 s) was used to load the samples onto the capillary and the detection wavelength was set at 191 nm.

The peak areas for the HAAs and TFAA as a function of concentration are shown in Table 5.1. Bromide was also detected in the samples and it was believed to be from an impurity in the neat chemicals or from HAA hydrolysis in the stock solutions. Nonlinear calibration curves were obtained for all samples (Table 5.1). It is unclear whether the HAAs interfered with each other, or whether the trace amounts of bromide interfered with HAA injection, or both.

The reproducibility of the electrokinetic injection method was evaluated by performing duplicate analyses of solutions containing 125, 250, and 500 µg/L of CAA. The buffer contained 12.5 mM NaH_2PO_4, 12.5 mM Na_2HPO_4, and 1 mM TTAOH at pH=5.7. A bubble cell (75 µm ID

Table 5.1
Peak areas of HAAs and TFAA as a function of concentration

Conc.	Peak Area (mAU•min.)									
(μg/L)	TFAA	CAA	BAA	DCAA	BCAA	DBAA	TCAA	BDCAA	CDBAA	TBAA
0	0.0	0.0	0.0	0.0	0.0	0.0	0.0	0.0	0.0	0.0
50	2.10	2.44	2.71	3.58	3.25	3.74	2.96	2.80	4.02	4.73
100	1.47	3.14	2.61	3.25	3.25	3.63	3.46	3.52	3.99	3.13
200	0.0	0.0	0.0	0.0	0.0	0.0	0.0	0.0	0.0	0.0
250	0.0	0.0	0.0	0.0	0.0	0.0	0.0	0.0	0.0	0.0
500	3.76	3.71	3.56	3.36	3.02	2.6	2.41	1.95	3.6	2.61
1000	35.3	54.1	58.4	72.4	79.2	56.3	71.4	50.7	51.0	47.2

Table 5.2
Peak areas of CAA as a function of concentration with duplicate analyses

	Peak area (mAU•min.)			
Conc. (μg/L)	Run 1	Run 2	Mean	RSD (%)
125	101.1	320.5	210.8	73.6
250	224.5	258.7	241.6	10.0
500	69.79	36.56	53.20	44.2

× 56 cm effective length) was used. Electrokinetic injection (–5 kV for 20 s) was used to load the samples onto the capillary and the detection wavelength was set at 191 nm.

The peak areas of CAA at different concentrations are shown in Table 5.2. There was a large difference between duplicate analyses with RSDs in the range of 10.0% to 73.6%. The calibration curve was also highly nonlinear (Figure 5.5). Furthermore, the peak area appeared to decrease with increasing concentration.

NDC Buffer

A 100 μg/mL stock solution of each HAA was prepared from 1 mg/mL stock solutions in MTBE (Supelco, Bellefonte, PA). These solutions were then used to prepare five composite calibration standards containing 2, 1, 0.667, 0.4, 0.2 mg/L of each HAA in DI water. TFAA was spiked into each composite standard at 100% of the HAA concentration. The buffer conditions for the separation were as follows: 4 mM NDC, 0.5 mM CTABr, pH=7.5. An electrokinetic injection of –100 kV•s was used and the detector was set at 235 nm for indirect detection. Linear calibration curves ($R^2 > 0.97$) were obtained for TFAA and HAAs spiked into DI water using electrokinetic injection (Figures 5.6 and 5.7).

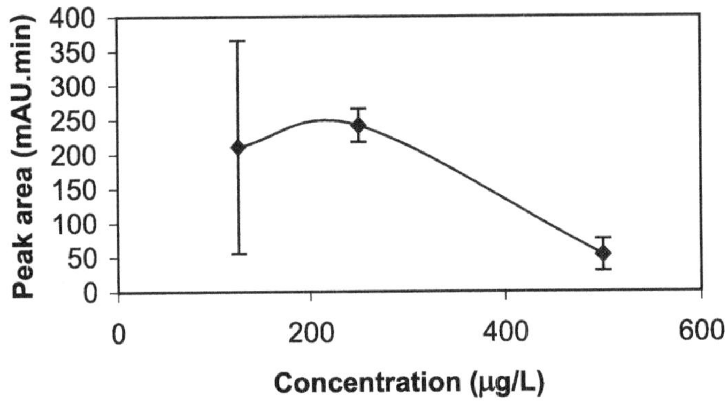

Figure 5.5 Calibration curve of CAA using electrokinetic injection

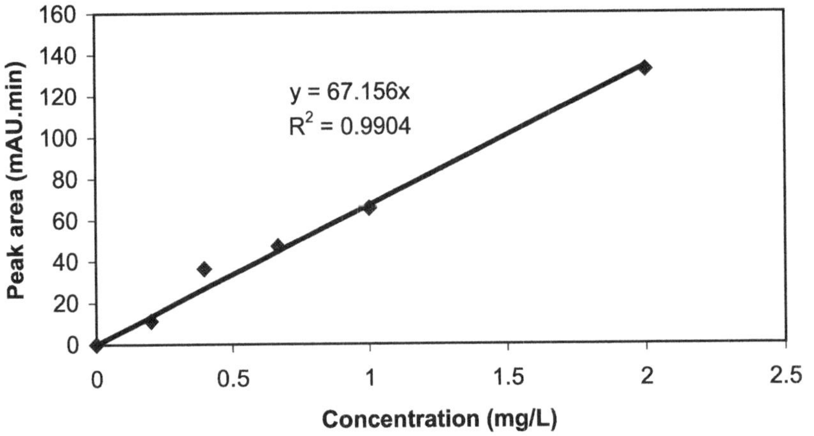

Figure 5.6 Calibration curve for TFAA using electrokinetic injection (–5 kV for 20 seconds)

EFFECT OF SAMPLE MATRIX ON ANALYSIS OF HAAs USING ELECTROKINETIC INJECTION

Phosphate Buffer

The effects of other solutes on HAA separation and detection using electrokinetic injection were also evaluated. The solutes tested included chloride, bromide and NOM. Suwanee River NOM was obtained from the International Humic Substances Society in powder form and served as a representative NOM source.

A composite solution of 1 mg/L of each of the nine HAAs and 1 mg/L TFAA was prepared by combining aliquots of aqueous stock solutions and diluting with DI water. Another composite solution of 1 mg/L of each of the nine HAAs and 1 mg/L TFAA together with 0.25 mg/L NOM as C, 0.5 mg/L NaCl, and 0.5 mg/L NaBr was prepared in the same manner.

For the analyses, the buffer contained 12.5 mM NaH_2PO_4, 12.5 mM Na_2HPO_4 and 1 mM TTAOH at pH=5.7. A straight capillary (75 µm ID × 56 cm effective length) was used.

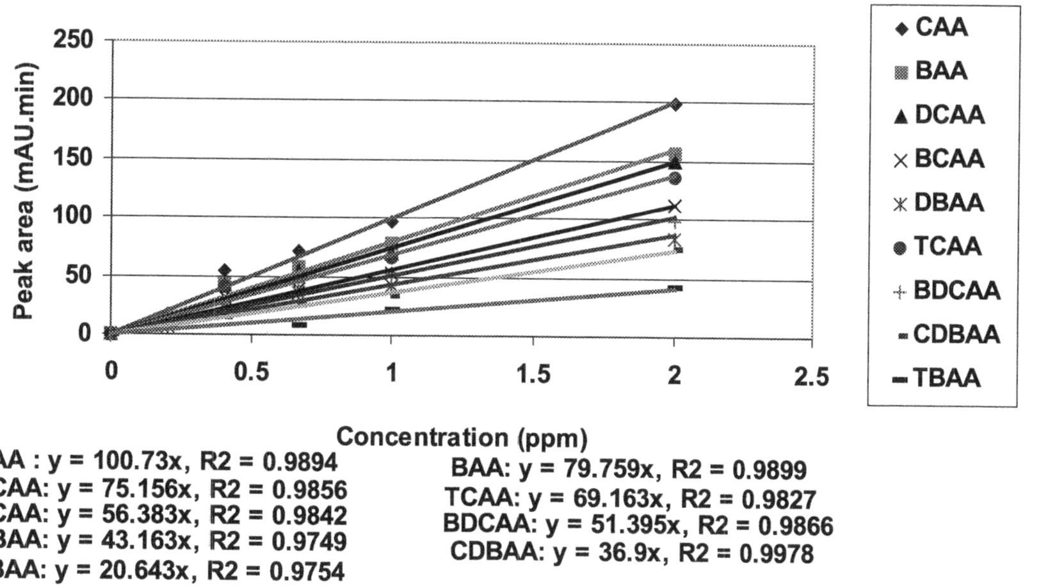

Figure 5.7 Calibration curves for the nine HAAs using electrokinetic injection (–5kV for 20 seconds)

Electrokinetic injection (–5 kV for 20 s) was used to load the samples onto the capillary and the detection wavelength was set at 191 nm.

With electrokinetic injection, the HAA peak areas decreased significantly when other solutes NOM, Cl^-, and Br^- were added to the sample (Figure 5.8 and Table 5.3). This is due the fact small and highly mobile anions like Cl^- and Br^- are preferentially loaded onto the capillary at the expense of the HAAs. Although it is expected that the organic molecules in the NOM sample would be predominantly larger than the HAA molecules, some of these macromolecules could have similar mobilities as HAAs if they have multiple anionic functional groups. Thus, they could potentially compete with the HAAs during sample loading. Also, it is possible that the NOM sample contains some small inorganic anions as the ash content of the NOM is 7% by weight. Therefore, sample matrix interferences may create problems for the use of electrokinetic injection for analysis of HAAs in finished water samples.

NDC Buffer

The ability to separate and detect HAAs spiked into a natural treated water sample following electrokinetic injection was also evaluated. The water sample consisted of raw water from St. Paul Regional Water Services (SPRWS) that was spiked with 1 mg/L of each of the nine HAAs and 1 mg/L of TFAA. The tail of a large peak from a low molecular weight anion with a migration time of approximately 4.2 minutes interfered with the HAA sample peaks (migration times of 4.4 to 5.7 minutes) (Figure 5.9). Thus, it appears that matrix effects may create problems for use of the electrokinetic injection method for analysis of drinking water samples.

The effect of a simple salt on the ability to separate and detect HAAs spiked into DI water following electrokinetic injection was evaluated. Three solutions of sodium chloride (0.001 M, 0.01 M, and 0.1 M) were prepared by dissolving NaCl into DI water. The resulting solutions were

A

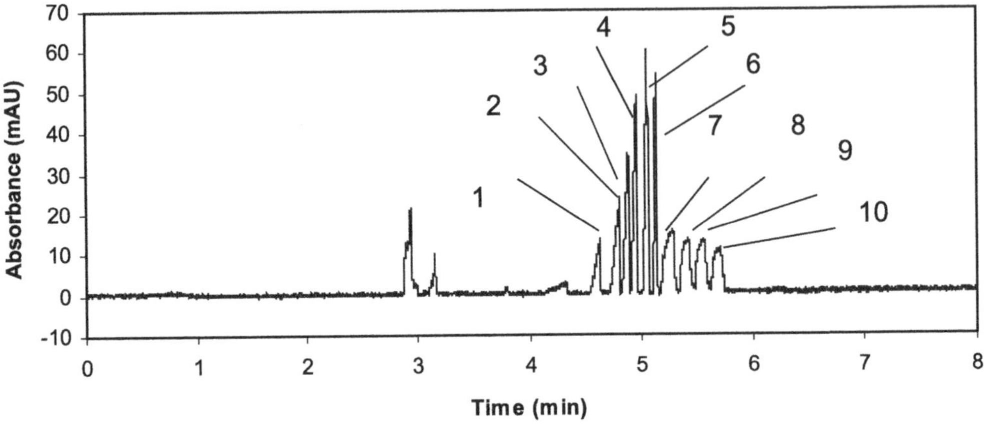

B

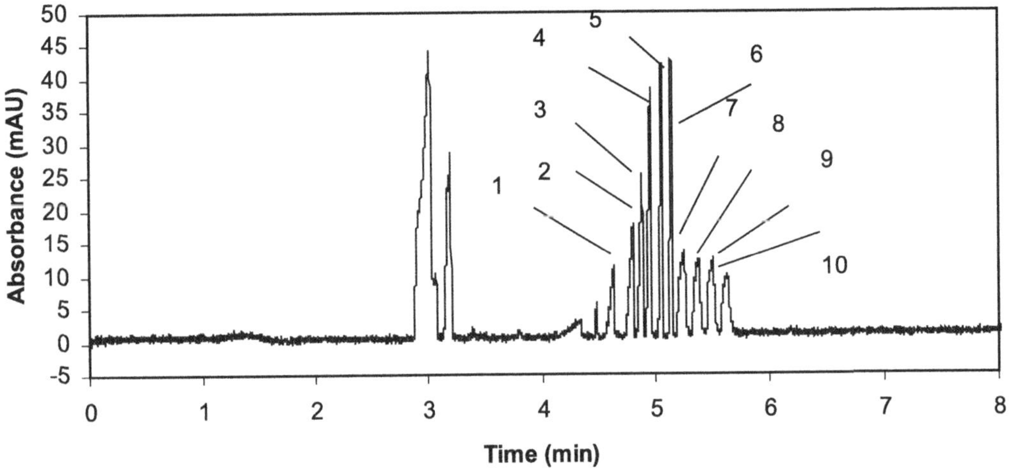

Figure 5.8 Evaluation of the effect of other solutes on the separation and detection of HAAs using electrokinetic injection
(A = 1 mg/L of each HAA and TFAA only, B = 1 mg/L of each HAA and TFAA + 0.5 mg/L NaCl + 0.5 mg/L NaBr + 0.25 mg/L NOM as C; Peak identification: 1-TFAA, 2-CAA, 3-BAA, 4-DCAA, 5-BCAA, 6-DBAA, 7-TCAA, 8-BDCAA, 9-CDBAA, 10-TBAA; Buffer composition: 12.5 mM NaH_2PO_4, 12.5 mM Na_2HPO_4, 1 mM TTAOH, pH = 5.7; Run conditions: Temperature = 25°C, Voltage = –20 kV, Injection = electrokinetic (–5 kV for 20 sec), Detection = direct UV at 191 nm; Capillary: straight capillary (75 μm ID 56 cm effective length))

Table 5.3
Effect of other solutes on HAA peak areas obtained using electrokinetic injection

	Peak area (mAU·min.)									
Sample	TFAA	CAA	BAA	DCAA	BCAA	DBAA	TCAA	BDCAA	CDBAA	TBAA
A	41.2	63.2	68.2	88.9	98.7	72.3	83.5	57.4	60.3	54.6
B	30.0	44.4	45.6	68.6	64.5	52.5	53.8	43.5	41.8	38.4
C	26.8	38.4	40.4	60.8	58.2	49.7	45.4	40.1	38.2	36.2

A — 1 mg/L of each HAA and 1 mg/L TFAA
B — 1 mg/L of each HAA and 1 mg/L TFAA with 0.25 mg/L NOM as C
C — 1 mg/L of each HAA and 1 mg/L TFAA with 0.25 mg/L NOM as C, 0.5 mg/L NaCl, and 0.5 mg/L NaBr

then spiked with 1 mg/L of each of the nine HAAs and 1 mg/L of TFAA. There was no overlap of the large chloride peak (migration time of 3.3 minutes) with the HAA peaks (Figure 5.10). Thus, it appears that chloride is not responsible for the matrix effects observed for the actual water sample. Nevertheless, increasing the salt concentration resulted in a decrease in peak area for CAA (Figure 5.11). Again, this is due to the fact a small and highly mobile anion like chloride is preferentially loaded onto the capillary at the expense of the HAAs.

CAPILLARY ISOTACHOPHORESIS

Selection of Leading and Terminating Electrolytes

The leading electrolyte must have a mobility greater than that of any of the components to be separated and the terminating electrolyte must have a mobility less than that of any of the analytes in the sample mixture. For anion analysis, chloride or nitrate are most frequently selected as the leading electrolyte due to their high mobilities. TBAA has the lowest mobility among the nine HAAs as it is the largest HAA and consistently the last HAA to exit the capillary. Therefore, the terminating electrolyte must have a mobility with a lower magnitude that of TBAA.

Materials and Methods

The electrophoretic mobilities of eight organic acids were measured by capillary electrophoresis by comparing the migration time of each compound to the migration time of a neutral marker compound. The buffer used was 12.5 mM phosphate with 5 mg/L hexadimethrine hydroxide (HDMOH) as the electroosmotic flow (EOF) modifier. HDMOH was prepared from the brominated form (HDMBr) by passing a solution of HDMBr through an Alltech Maxi-clean IC-OH plus ion exchange column. The pH of the buffer was adjusted to 5.7 with phosphoric acid. The buffer was filtered (0.2 μm membrane filter) and vacuum degassed prior to use. Solutions of organic acids (20 mg/L) were prepared by dissolving reagent grade chemical in distilled and deionized water. Mesityl oxide served as the neutral marker compound. A standard 75 μm ID capillary with a total length (L_{tot}) of 64 cm and an effective length (L_d) of 58 cm was used. The

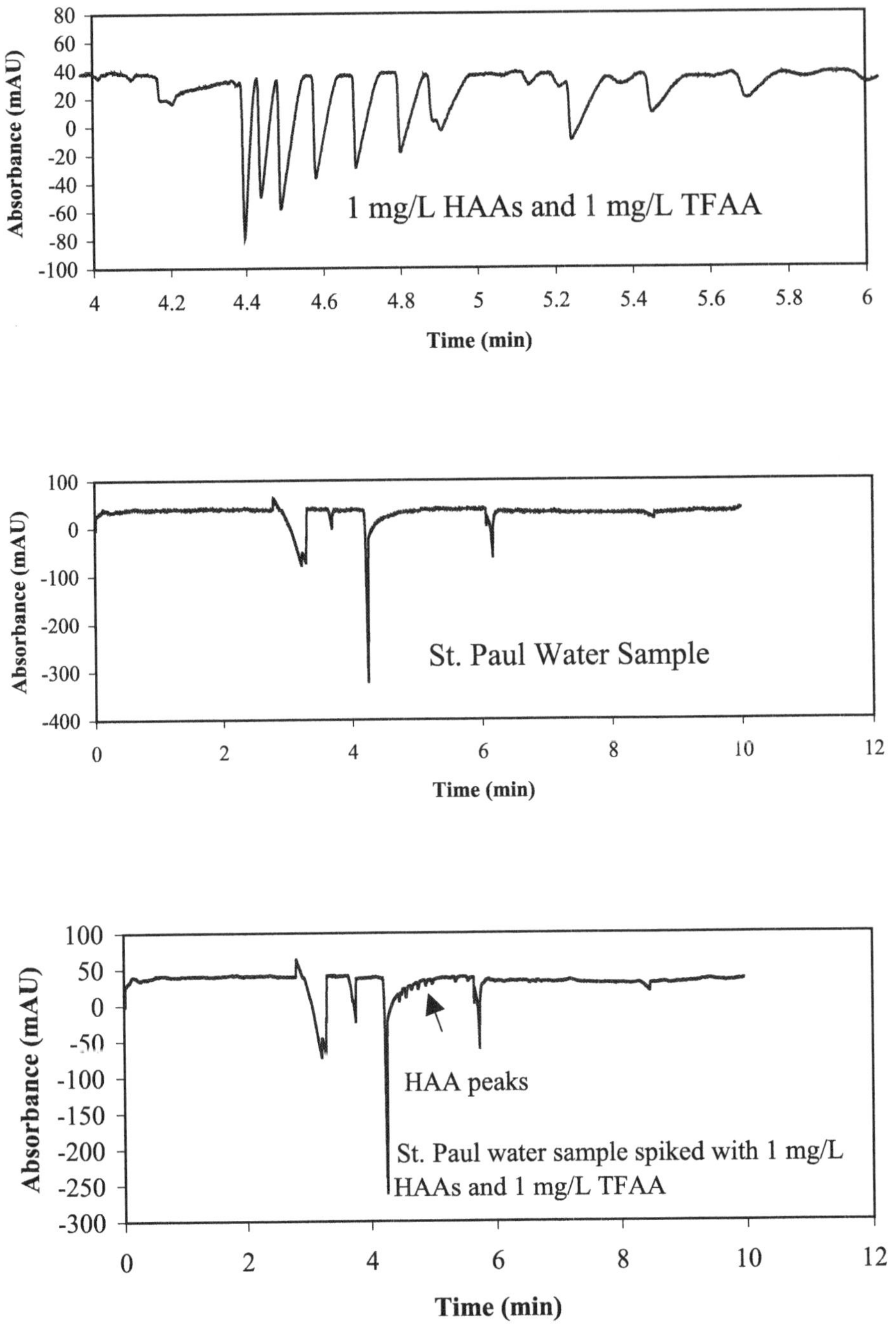

Figure 5.9 Effect of matrix constituents in an SPRWS water sample on the analysis of HAAs using electrokinetic injection
(Buffer composition: 4 mM NDC, 1 mM CTABr, pH = 7.5; Run conditions: Temperature = 25°C, Voltage = –20 kV, Injection = electrokinetic (–100 kV·s), Detection = indirect UV at 235 nm; Capillary: straight capillary (75 μm ID × 56 cm effective length))

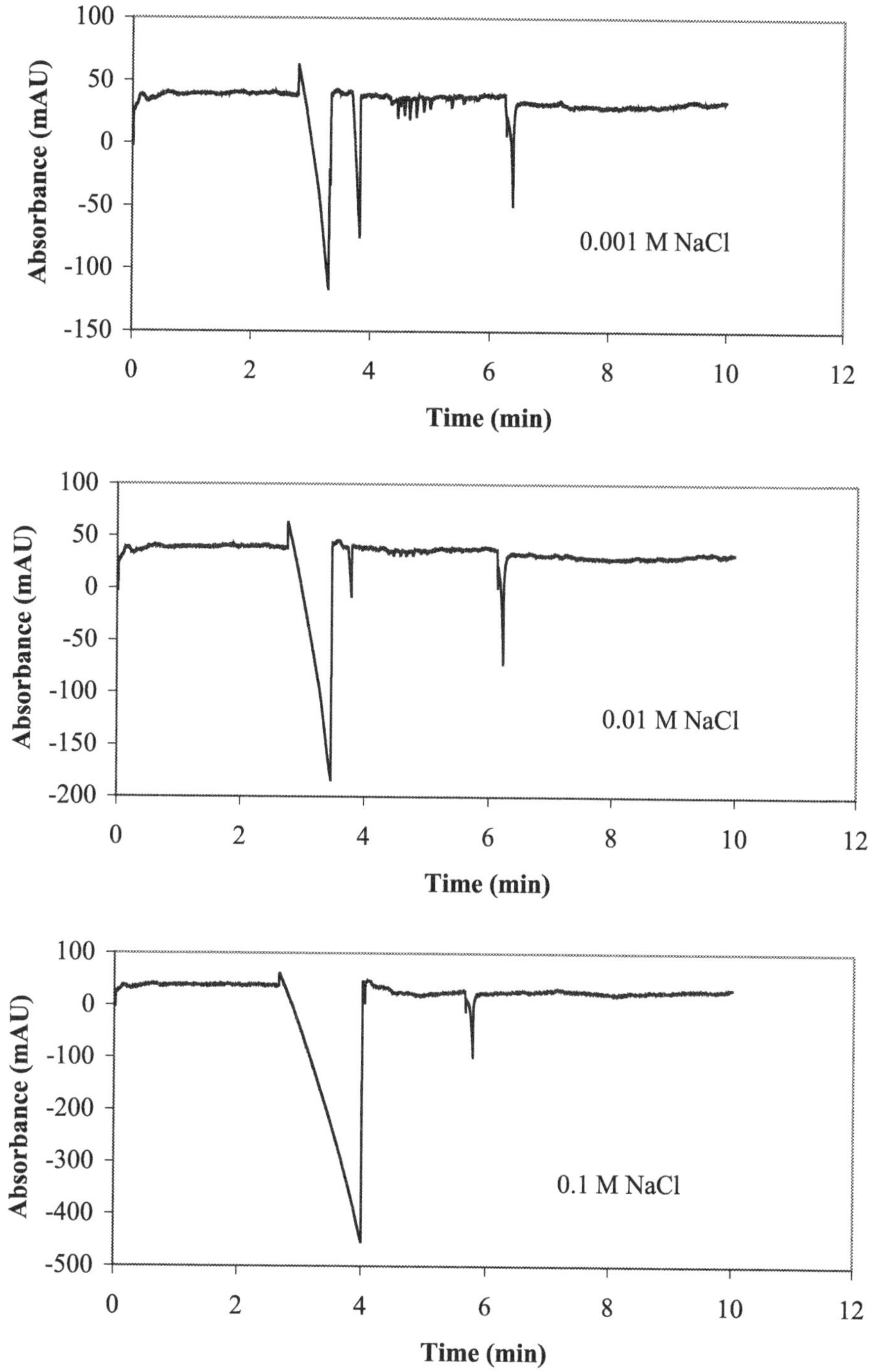

Figure 5.10 Effect of NaCl concentration on the analysis of HAAs using electrokinetic injection

(Buffer composition: 4 mM NDC, 1 mM CTABr, pH = 7.5; Run conditions: Temperature = 25°C, Voltage = –20 kV, Injection = electrokinetic (–100 kV·s), Detection = indirect UV at 235 nm; Capillary: straight capillary (75 μm ID × 56 cm effective length))

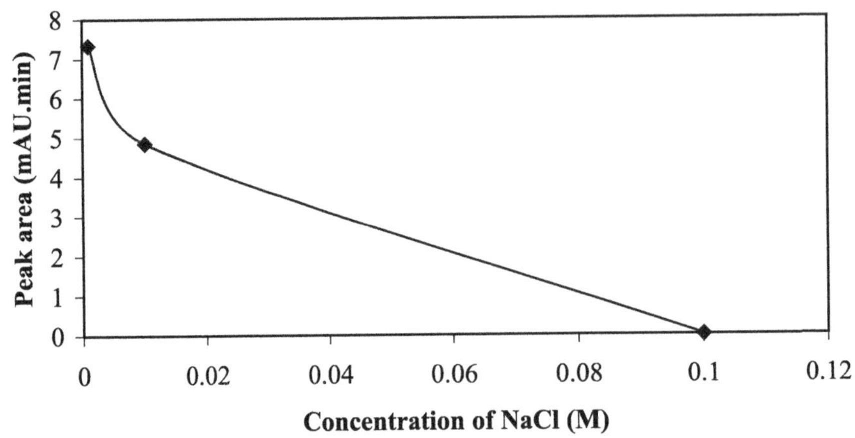

Figure 5.11 Effect of NaCl concentration on CAA peak area obtained using electrokinetic injection

applied voltage was –20 kV. The detection wavelength for all of the compounds was 191 nm. The following formula was used to calculate the electrophoretic mobilities (EM) (Glynn et al. 1998):

$$EM = (L_{tot}L_d/V)(1/t_p - 1/t_{ref}) \tag{5.1}$$

where L_d = distance from the capillary inlet to the detector, L_{tot} = total length of the capillary, V = applied voltage, t_{ref} = mean migration time of the reference marker, and t_p = mean migration time of the analyte.

Results and Discussion

The results from the evaluation of electrophoretic mobilities are shown in Table 5.4. The mobility of tribromoacetic acid was -3.24×10^{-8} $m^2V^{-1}s^{-1}$ and only those compounds with mobilities of lower magnitude were capable of serving as the terminating electrolyte for CITP. All of the organic acids tested except for EDTA had mobilities that were lower in magnitude than the value for TBAA and were considered as possible terminating electrolytes for future study.

Evaluation of On-Capillary Transient CITP-CZE

Materials and Methods

The background electrolyte consisted of 12.5 mM phosphate with 1 mM TTAOH as EOF modifier. The buffer was adjusted to pH 5.7 with phosphoric acid. The leading electrolyte (KNO_3) was mixed together with the sample and contained 0.01 M KNO_3, 5 mg/L of each of the nine HAAs, and 5 mg/L of TFAA as an internal standard. This mixture was injected onto the capillary and then 0.02 M 1-L-glutamic acid was injected as the terminating electrolyte. A constant current of (–22 μA) was applied for 2 minutes for CITP migration. Then, the terminating electrolyte was replaced by the background electrolyte and constant current or constant voltage was applied for the capillary electrophoresis separation. Different injection times and different current/voltage

Table 5.4
Ranking of the electrophoretic mobilities of TBAA and eight organic acids from highest magnitude to lowest magnitude

	t_p (min)	t_{ref} (min)	Mobility ($m^2V^{-1}s^{-1}$)
EDTA	4.266	9.340	-3.94×10^{-8}
TBAA	4.710	9.301	-3.24×10^{-8}
Benzoic acid	4.782	9.280	-3.14×10^{-8}
Phenoxyacetic acid	4.928	9.182	-2.91×10^{-8}
3-hydroxybenzoic acid	4.972	9.253	-2.88×10^{-8}
3,5-dichlorobenzoic acid	5.022	9.201	-2.80×10^{-8}
4-ethylbenzoic acid	5.122	9.211	-2.68×10^{-8}
1-L-glutamic acid	5.163	9.150	-2.61×10^{-8}
2,4-dichlorophenoxyacetic acid	5.229	9.338	-2.60×10^{-8}

Table 5.5
CITP operating conditions

Injection time for the sample containing the leading electrolyte	Injection time for the terminating electrolyte	Current for CITP	Current or voltage for CZE separation
50.0 mbar for 30 s	50.0 mbar for 4 s	22 μA	22 μA
50.0 mbar for 4 s	50.0 mbar for 1 s	22 μA	22 μA
50.0 mbar for 2 s	50.0 mbar for 1 s	22 μA	22 μA
50.0 mbar for 1 s	50.0 mbar for 1 s	22 μA	22 μA
50.0 mbar for 0.5 s	50.0 mbar for 0.5 s	22 μA	22 μA
50.0 mbar for 0.5 s	50.0 mbar for 0.5 s	22 μA	40 μA
50.0 mbar for 0.5 s	50.0 mbar for 0.5 s	22 μA	20 kV

settings were tested to evaluate the effects of these variables on the preconcentration and separation of HAAs and TFAA (Table 5.5).

Results and Discussion

An example of the on-capillary transient CITP sample preconcentration followed by CZE separation of HAAs is shown in Figure 5.12. For comparison, an example of a simple CZE separation of HAAs is shown in Figure 5.13. There appeared to be a sample preconcentration or stacking effect from CITP as seen by comparing the height of the HAA band in the CITP-CZE run (400 to 650 mAU; Figure 5.12) with the HAA peak heights in the CZE only run (2 to 8 mAU;

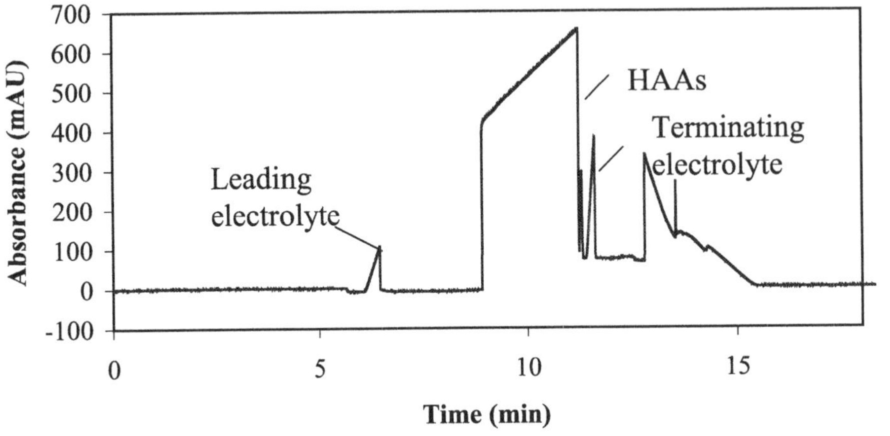

Figure 5.12 On-capillary transient CITP-CZE analysis of a standard solution of nine HAAs and TFAA
(Leading electrolyte: 0.01 M KNO_3; Terminating electrolyte: 0.02 M 1-L-glutamic acid; Capillary: standard capillary (75 µm ID × 58 cm effective length); Sample: 5 mg/L HAAs and 5 mg/L TFAA dissolved in the leading electrolyte solution; Injections: 50 mbar for 4 seconds (leading electrolyte + sample) and 50 mbar for 1 second (terminating electrolyte); Current: held constant at –22 µA)

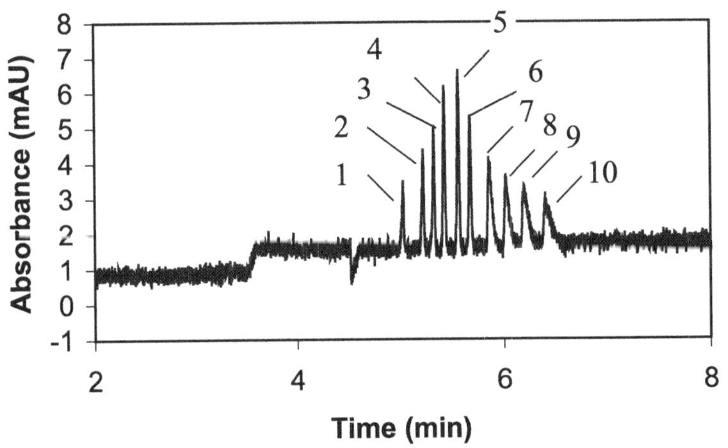

Figure 5.13 CZE separation of a standard solution of nine HAAs and TFAA
(Sample: 10 mg/L HAAs and 10 mg/L TFAA; Buffer: 25 mM phosphate, 1 TTAOH, pH = 5.7; Capillary: standard capillary (75 µm ID × 58 cm effective length); Injection: pressure 50 mbar for 4 seconds; Voltage: –20 kV)

Table 5.6
Characteristics of LiChrolut EN columns

Sorbent	Non-ionogenic, highly-porous polystyrene-divinylbenzene polymer
Particle distribution	40-120 μm
Specific surface	1200 m^2/g
Pore volume	0.75 mL/g
pH value	Neutral
Stability	pH 1 to 13
Capacity	500 mg Caffeine/g sorbent (model substance for polar analytes)
	500 mg Diisodecylphthalate (DIDP)/g sorbent (model substance for non-polar analytes)

Figure 5.13). Unfortunately, the HAA sample peaks were unresolved under all conditions. Based on our limited but unsuccessful experience with CITP, we decided to abandon this method in favor of off-line solid-phase extraction.

OFF-LINE SOLID-PHASE EXTRACTION

Materials and Methods

LiChrolut EN columns (EM Science, Gibbstown, NJ, catalog # 19870-1) were used to sorb and concentrate the HAAs at low pH. The characteristics of the columns are shown in Table 5.6. Due to its large accessible specific surface area and slightly hydrophilic character, LiChrolut EN columns are suitable for retention and recovery of polar analytes (Loos and Niessner 1998, 1999).

The preliminary SPE process involved loading the aqueous sample onto the preconditioned (methanol followed by water) column, drying the column by application of a vacuum, elution of the HAAs with methanol and dilution of the extract with DI water prior to analysis by CE. The SPE process was optimized by some preliminary experiments in which the sample consisted of 5 mg/L of each HAA dissolved in DI water. In these experiments, the effects of sample pH, sample loading rate, column rinse conditions, volume of methanol used for HAA elution from the column, and sample volume were evaluated.

All of the glassware (e.g., beakers, test tubes, glass funnels, glass adaptors, volumetric flasks) was cleaned according to the following procedure before use: 1) soaked overnight in 3 N nitric acid; 2) rinsed 3 to 4 times with distilled water; 3) rinsed with Milli-Q water (Millipore Corp., MA); and 4) baked at 550°C for 6 hours.

A 9-port reati-vap evaporator (Pierce, Rockford, IL) primarily comprised of stainless steel was used to evaporate the methanol from the SPE concentrate prior to injection. A capillary grade hydrocarbon trap (Chrom tech, Apple Valley, MN) was installed between the nitrogen tank and the evaporator to remove organic contaminants from the nitrogen stream. In addition, the evaporator was thoroughly cleaned before use. Water, methanol, acetone, and hexane were pumped through the evaporator and stainless steel tubing sequentially to clean the evaporator. Finally, the

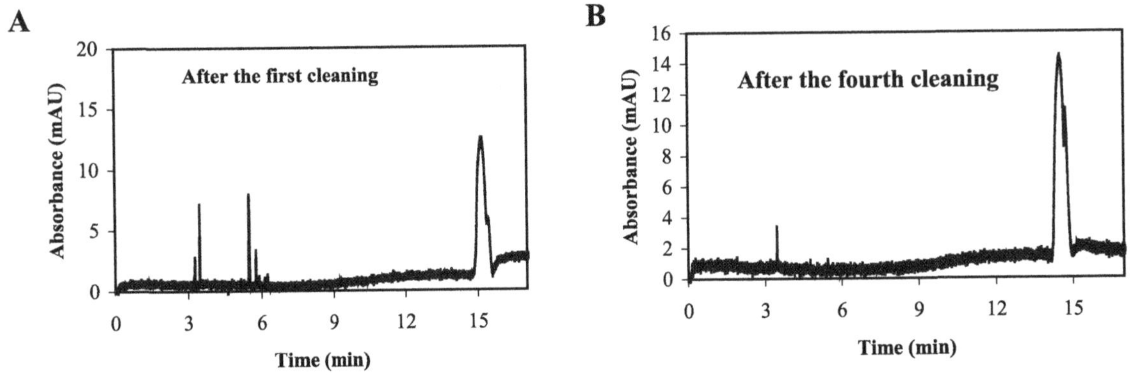

Figure 5.14 Contamination peaks observed in first cleaning (A) and fourth cleaning (B) of the evaporator
(Buffer composition: 12.5 mM NaH_2PO_4, 12.5 mM Na_2HPO_4, 1 mM TTAOH, pH = 5.7; Run conditions: Temperature = 25° C, Voltage = –20 kV, Injection = pressure (200 mbar·s), Detection = direct UV at 191 nm, Capillary: bubble cell (75 μm ID × 56 cm effective length))

evaporator and its stainless steel tubing were dried using a nitrogen stream. The cleaning procedure was repeated four times to reduce the potential for contaminant introduction by the methanol evaporation step.

A simple experiment was used to test the cleanliness of the evaporation system. In this experiment, 1 mL of methanol (pesticide residue analysis grade, Burdick-Jackson, Muskegon, MI) was transferred to a clean glass test tube directly from the methanol squeeze bottle (comprised of PTFE). Then, 0.5 mL of Milli-Q water was added to the test tube. The methanol-H_2O solution was mixed using a vortex mixer. Finally the solution was blown down to 0.5 mL using the evaporator and the resulting solution was subsequently analyzed using CE. The electropherograms of this 0.5 mL "blank" solution after the first cleaning and the fourth cleaning are shown in Figure 5.14. There was a clear reduction in the number of contaminant peaks from the first cleaning to the fourth and only one peak appeared in the electropherogram (at 3.4 minutes) after the fourth and final cleaning. A peak at 3.4 minutes was observed when a sample of methanol was analyzed directly (results not shown). Thus, it was concluded that this contamination peak at 3.4 minutes resulted from the methanol and that the evaporator and its stainless steel tubing were clean. Fortunately, the 3.4-minute contamination peak does not interfere with the analysis of HAAs as the migration times of HAAs are in the range of 5 to 7 minutes.

Sulfuric acid with purity of 99.999% (cat#: 33974-1, Sigma-Aldrich, Milwaukee, WI) was used to adjust the pH of the samples. Pesticide residual analysis grade methanol (Burdick & Jackson, Muskegon, MI) was used in the preconditioning and elution step. Bottled high performance liquid chromatography (HPLC) grade water was used to prepare the HAAs standard solutions and blank solutions.

Results

Method Optimization

The preliminary results indicated that the sample pH significantly affected HAA recovery. High HAAs recoveries were obtained (43% for CAA; 76 to 90% for the other HAAs) when the sample was acidified to pH 2, however, the recoveries were much lower at pH 3.5 (3 to 20%) and 6.86 (0 to 3%). The HAAs were more hydrophobic and recoveries increased when the sample pH was decreased to near or below the pKa values of the compounds which range from 0.51 to 2.92 (Serjeant and Dempsey 1979). Lower sample pH values (i.e., pH < 2) were not tested because the lowest value recommended by the manufacturer to ensure stability of the LiChrolut EN columns is pH 1 (Table 5.6).

Sample loading rate was another important factor in HAA recovery. Recoveries of HAAs were generally lower and more variable (37% for CAA; 56 to 97% for the other HAAs) when the sample loading rate was not controlled (≥5 mL/min). However, recoveries of HAAs were higher (42% for CAA; 73 to 90% for the other HAAs) and more reproducible when the sample loading rate was controlled at 2 mL/min. Further reduction in the sample loading rate to 1 mL/min improved the recovery of CAA and that loading rate was selected for the final method.

In general, recoveries of HAAs decreased with increasing rinse water volume. When 2 mL of pH 2 water was used to rinse the column, the recoveries were 40% for CAA and 70 to 91% for the other HAAs. Decreasing the rinse water volume to 0.5 mL increased the recoveries to 43% for CAA and 76 to 91% for the other HAAs. The recoveries also decreased when the pH of the rinse water was increased to 5.6. Thus, 0.5 mL of HPLC grade water adjusted to pH 2 with sulfuric acid was selected for the final method.

Sample volume was another important factor, with higher HAA recoveries observed for decreasing sample volumes. For example, when a 100 mL sample containing 50 µg/L of each HAA was processed, the HAA recoveries were less than 82.5% for all but one compound (BDCAA). However, the recoveries were greater than 90% for all compounds except CAA and BDCAA when only 50 mL of sample was processed.

Additional experiments were performed to evaluate the effect of the methanol elution volume on HAA recoveries. In these experiments, the column drying step was not used and the sample consisted of 50 mL of 50 µg/L HAAs in DI water. Also, the diluted methanol extract was blown down to a final volume of 0.5 mL. When the methanol volume used to elute the HAAs was 1.3 mL, the recoveries were 59% for CAA and 46 to 97% for the other HAAs. The recoveries improved to 63% for CAA and 74 to 98% for the other HAAs when the volume of methanol was increased to 2.3 mL. Higher volumes of methanol were not used for two reasons: 1) the recoveries for 2.3 mL were acceptable and 2) the time required for the methanol evaporation or blow down step is proportional to the methanol volume used.

The final SPE process using preconditioned LiChrolut EN columns is shown in Figure 5.15: 1) acidification of the sample to pH 2, 2) sample loading at 1 mL/min, 3) column rinsing with 0.5 mL of pH 2 HPLC grade water (Mallickrodt Baker, Paris, KY), 4) elution of HAAs with 2.3 mL of methanol, 5) dilution with 0.2 mL of HPLC grade water, and 6) evaporation of methanol (blow down) prior to analysis by CE. The LiChrolut EN columns were preconditioned with 30 mL of methanol followed by 10 mL of HPLC grade water at pH 2. The extraction process was carried out using the Visiprep SPE vacuum system (SupelCo, Bellefonte,

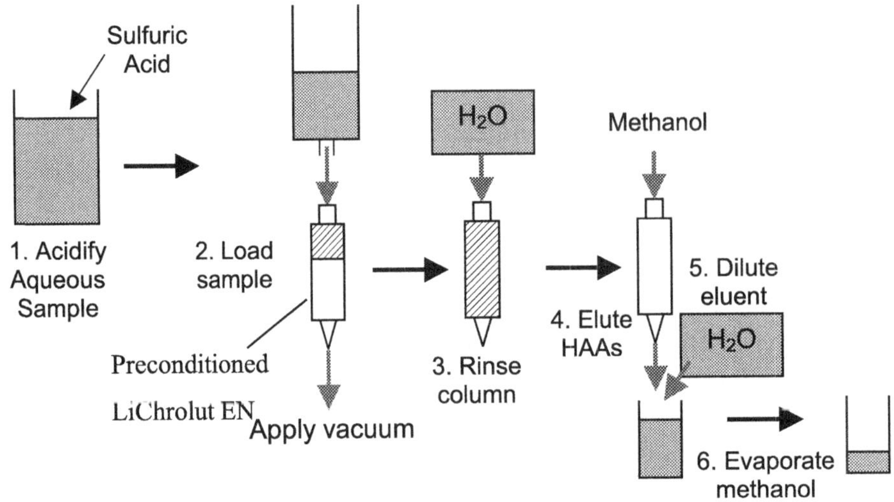

Figure 5.15 SPE using LiChrolut EN columns

PA). The column drying process was eliminated from the final procedure because of the concern over contamination of the column by chemicals in the room air.

Determination of HAA Recoveries and MDLs for the Final SPE Method

Due to its relatively high volatility, TFAA was eliminated as the internal standard and external calibration was used to quantify all of the peaks. Selecting a compound as surrogate for the SPE process is important for monitoring sample extraction efficiency. Ideally, the surrogate will have similar physical and chemical properties as the analytes and is not typically present in the water samples to be tested. Triiodoacetic acid, bromodiiodoacetic acid, or dibromoiodoacetic acid were considered for use as the surrogate, however, they are not stable and not commercially available. Other compounds such as 2,2,3-tribromopropanoic acid, 2,3,3-tribromopropanoic acid, 2,2,3,3-tetrabromopropanoic acid, 2,3,3,3-tetrabromopropanoic acid, or pentabromopropanoic acid were also considered; however, those compounds are not commercially available either. Because TBAA has similar physical-chemical properties as the HAA_5 compounds and it is rarely detected in drinking water (Krasner et al. 2000), it was selected as the surrogate in the SPE process.

Solutions consisting of 50 mL or 25 mL of 50 µg/L HAAs and 200 µg/L TBAA prepared from aqueous standards were used to determine the HAA recoveries using the SPE process. The final volume of the concentrated solution after SPE and methanol evaporation was 0.5 mL. Electropherograms from the analysis of the final 0.5 mL of solution are shown in Figures 5.16 and 5.17. An electrophoregram from the analysis of the final 0.5 mL of solution from a blank run (DI water sample) is shown in Figure 5.18. In general, the peak resolution and peak shapes were excellent for the SPE-preconcentrated sample (Figures 5.16 and 5.17). Nevertheless, a contamination peak at 6.8 minutes comigrated with the CDBAA peak. Thus, CDBAA can't be quantified accurately with this method. The source of this contamination peak is believed to be the SPE column itself.

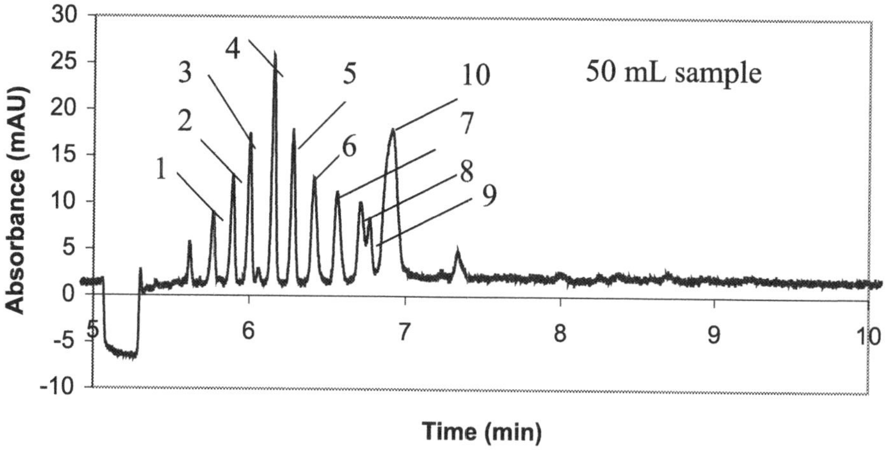

Figure 5.16 CE electropherogram from the analysis of a 0.5 mL extract from the SPE preconcentration of a 50 mL sample containing 50 µg/L HAAs and 200 µg/L TBAA (Peak identification: 1-CAA, 2-BAA, 3-DCAA, 4-BCAA, 5-DBAA, 6-TCAA, 7-BDCAA, 8-contamination peak, 9-CDBAA, 10-TBAA; Buffer composition: 12.5 mM NaH_2PO_4, 12.5 mM Na_2HPO_4, 1 mM TTAOH, pH = 5.7; Run conditions: Temperature = 25°C, Voltage = –20 kV, Injection = pressure (600 mbar·s), Detection = direct UV at 191 nm; Capillary: bubble cell (75 µm ID × 56 cm effective length))

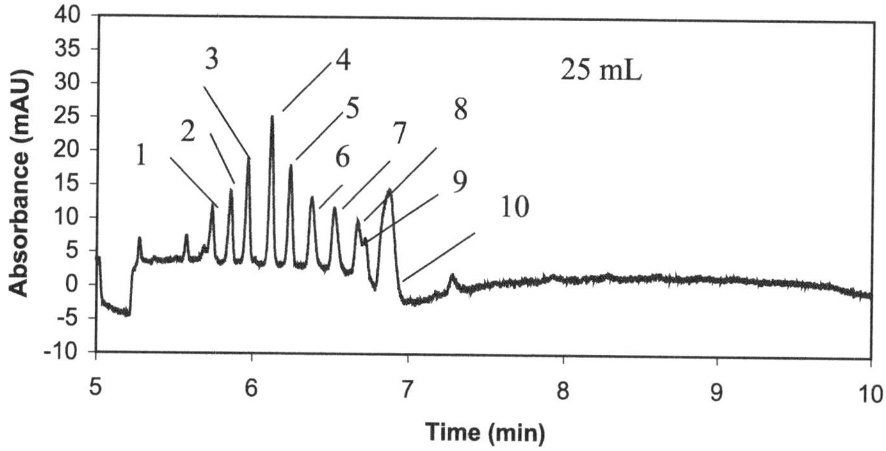

Figure 5.17 CE electropherogram from the analysis of a 0.5 mL extract from the SPE preconcentration of a 25 mL sample containing 50 µg/L HAAs and 200 µg/L TBAA (Peak identification: 1-CAA, 2-BAA, 3-DCAA, 4-BCAA, 5-DBAA, 6-TCAA, 7-BDCAA, 8-contamination peak, 9-CDBAA, 10-TBAA; Buffer composition: 12.5 mM NaH_2PO_4, 12.5 mM Na_2HPO_4, 1 mM TTAOH, pH = 5.7; Run conditions: Temperature = 25°C, Voltage = –20 kV, Injection = pressure (600 mbar·s), Detection = direct UV at 191 nm; Capillary: bubble cell (75 µm ID × 56 cm effective length))

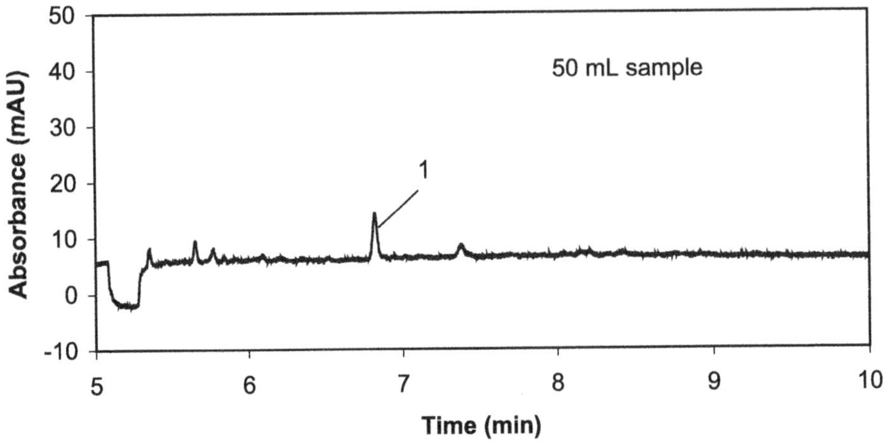

Figure 5.18 CE electropherogram from the analysis of a 0.5 mL extract from the SPE preconcentration of a 50 mL blank sample
(Buffer composition: 12.5 mM NaH_2PO_4, 12.5 mM Na_2HPO_4, 1 mM TTAOH, pH = 5.7; Run conditions: Temperature = 25°C, Voltage = –20 kV, Injection = pressure (600 mbar·s), Detection = direct UV at 191 nm; Capillary: bubble cell (75 μm ID × 56 cm effective length))

The HAA recovery values are shown in Table 5.7. When 50 mL of sample was processed, the recoveries for HAAs and TBAA were in the range of 63% to 99%. Except for CAA, the HAA recoveries were in the range of 75% to 99%. These recoveries are suitable for quantitative analysis. The low recovery for CAA (63%) was attributed to its relatively hydrophilic character and may be deemed too low for quantification. If a higher CAA recovery is desired, then the sample volume can be reduced to 25 mL where a CAA recovery of 81% was obtained (Table 5.7). Therefore, two runs may be needed to quantify HAAs. One run using 25 mL of sample to quantify CAA and the other run using 50 mL of sample to quantify the other HAAs.

The MDL (without SPE) was divided by the theoretical concentration factor to obtain the overall MDL. The theoretical increase in sensitivity is 100 (i.e., 50 mL/0.5mL) when 50 mL of sample is processed and 50 when 25 mL of sample is processed. The overall MDL for CAA was 5 μg/L and the MDLs for the other HAAs ranged from 1.7 to 2.5 μg/L (Table 5.7).

SUMMARY

Electrokinetic injection and capillary isotachophoresis as the on-capillary techniques and solid-phase extraction as the off-line technique were evaluated as potential HAA preconcentration methods. Significant improvements in sensitivity were obtained when electrokinetic injection was used to load the analytes onto the capillary. Unfortunately, the non-linearity of the calibration curves and a lack of reproducibility were observed, especially when the sample matrix contained anions other than HAAs alone. These effects may be problematic for routine analysis of HAAs in drinking water. It appeared that sample stacking was obtained during the CITP-CZE runs. Unfortunately, the HAA sample peaks were not able to be resolved under any of the conditions tested. SPE using LiChrolut EN columns was successful at enriching all nine HAAs. The

Table 5.7
Recoveries and MDLs for the nine HAAs using the SPE-CE method

HAA	Mean recovery (%) for 50 mL of sample	Mean recovery (%) for 25 mL of sample	MDL without SPE (μg/L)	MDL with SPE (50 mL) (μg/L)	MDL with SPE (25 mL) (μg/L)
CAA	63 ± 7.6	81 ± 8.5	250	2.5	5.0
BAA	96 ± 2.3	91 ± 1.5	220	2.2	4.4
DCAA	93 ± 4.4	90 ± 6.9	180	1.8	3.6
BCAA	98 ± 3.3	92 ± 6.5	170	1.7	3.4
DBAA	96 ± 1.4	92 ± 3.6	220	2.2	4.4
TCAA	95 ± 1.6	93 ± 4.9	200	2.0	4.0
BDCAA	99 ± 1.5	97 ± 9.8	210	2.1	4.2
CDBAA	85 ± 2.5	75 ± 5.8	230	2.3	4.6
TBAA	74 ± 1.5	80 ± 6.2	240	2.4	4.8

recoveries were in the range of 75% to 99% for all HAAs except CAA (63%) when 50 mL of sample was passed through the column. The recovery for CAA increased to 81% when the sample was reduced to 25 mL. The overall MDLs including SPE preconcentration and analysis by CE ranged from 1.7 to 2.5 μg/L for all HAAs except CAA (5 μg/L).

CHAPTER 6
FINAL METHOD REFINEMENT, TESTING AND VERIFICATION

EFFECT OF OTHER SOLUTES ON HAA SEPARATION AND DETECTION BY CE

Introduction

The effects of chloride, bromide, and NOM on HAA separation and detection were evaluated to determine whether matrix effects would present or limit routine use of the CE-based method for analysis of drinking water samples. Suwanee River NOM was obtained from the International Humic Substances Society in powder form and was used as a representative NOM source.

Materials and Methods

A composite standard solution containing 10 mg/L of each of the nine HAAs and 10 mg/L TFAA was prepared by combining aliquots of aqueous stock solutions and diluting with DI water. Another composite solution of 10 mg/L of each of the nine HAAs and 10 mg/L TFAA together with 2.5 mg/L NOM as carbon, 5 mg/L NaCl, and 5 mg/L NaBr was prepared in the same manner. The two composite solutions were analyzed by CE using a straight capillary (75 μm ID × 56 cm effective length) and the phosphate-TTAOH buffer. Pressure injection (200 mbar•s) was used to load the samples onto the capillary and the detection wavelength was set at 191 nm.

Results

No significant changes in peak shape and peak migration time were observed using pressure injection when the other solutes were added (Figure 6.1). The HAA peak areas from the analysis of various solutions are compared in Table 6.1. With pressure injection, the peak areas of the HAAs and TFAA in the presence of NOM, NaCl, and NaBr were similar to those in the absence of these solutes, suggesting that matrix effects are not important when pressure injection is used.

EVALUATION OF THE EFFECT OF NOM ON HAA RECOVERIES USING SPE PROCESS

Introduction

NOM is ubiquitous in drinking water supplies. According to a survey of U.S. water utilities by Kaplan (2000), dissolved organic carbon (DOC) concentrations in drinking water supplies ranged from 0.23 mg/L for a groundwater source to 38.01 mg/L for a surface water source from the Atlantic coastal plain. The median concentration for all systems surveyed was 3.13 mg/L. Assuming 50% NOM removal during coagulation of surface waters, the DOC concentration in finished water may range from 0.23-19 mg/L. These NOM concentrations are approximately three orders-of-magnitude greater than typical HAA concentrations. NOM may interfere with the analysis of HAAs because NOM may occupy sorption sites on the SPE column and reduce HAA recoveries. Thus, the effect of

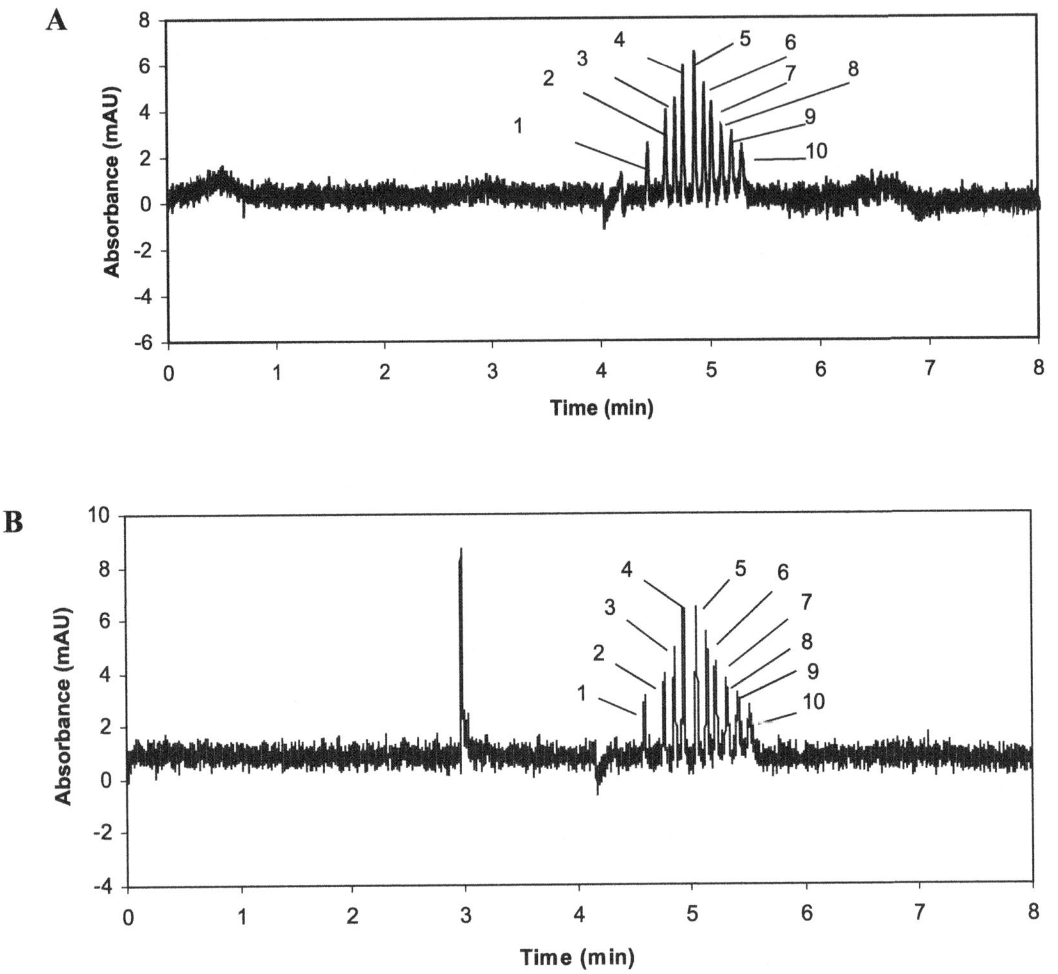

Figure 6.1 Evaluation of the effect of other solutes on the separation and detection of HAAs using pressure injection
(A = 10 mg/L HAAs and TFAA only, B = 10 mg/L HAAs and TFAA + 5 mg/L NaCl + 5 mg/L NaBr + 2.5 mg/L NOM as carbon; Peak identification: 1-TFAA, 2-CAA, 3-BAA, 4-DCAA, 5-BCAA, 6-DBAA, 7-TCAA, 8-BDCAA, 9-CDBAA, 10-TBAA; Buffer composition: 12.5 mM NaH_2PO_4, 12.5 mM Na_2HPO_4, 1 mM TTAOH, pH = 5.7; Run conditions: Temperature = 25°C, Voltage = –20 kV, Injection = pressure (200 mbar·s), Detection = direct UV at 191 nm; Capillary: straight capillary (75 μm ID × 56 cm effective length)

Table 6.1
Effect of other solutes on HAA peak areas using pressure injection (200 mbar·s)

HAA	Sample A	Sample B	Sample C
TFAA	3.03	2.63	2.87
CAA	4.70	4.17	4.26
BAA	5.31	5.16	5.07
DCAA	7.12	6.48	7.08
BCAA	8.56	7.67	7.29
DBAA	6.99	6.51	6.39
TCAA	6.07	5.89	6.11
BDCAA	4.07	4.53	4.39
CDBAA	4.36	4.52	4.14
TBAA	3.32	3.46	4.45

A — 10 mg/L of each HAA and TFAA
B — 10 mg/L of each HAA and TFAA with 2.5 mg/L NOM
C — 10 mg/L of each HAA and TFAA with 2.5 mg/L NOM, 5 mg/L NaCl, 5 mg/L NaBr

NOM on HAA recovery by SPE was evaluated using Suwanee River NOM obtained from the International Humic Substances Society as a representative NOM source.

Materials and Methods

The SPE process was used to preconcentrate the HAAs in the following solutions: 1) 50 μg/L of each HAA and 200 μg/L TBAA; 2) 50 μg/L of each HAA and 200 μg/L TBAA with 1 mg/L NOM as C; and 3) 50 ppb of each HAA and 200 ppb TBAA with 5 mg/L NOM as carbon. The original sample volume was 50 mL and the final 0.5 mL concentrated solutions were analyzed for HAAs using CE. For the analyses, the buffer contained 12.5 mM NaH_2PO_4, 12.5 mM Na_2HPO_4 and 1 mM TTAOH at pH=5.7. A bubble cell capillary (75 μm ID × 56 cm effective length) was used. Pressure injection (600 mbar·s) was used to load the samples onto the capillary and the detection wavelength was set at 191 nm. In addition, an experiment was performed to track the fate of the NOM during the SPE process by processing a 50 mL sample containing only 5 mg/L NOM as C and using UV absorbance at 254 nm to quantify the amount of NOM.

Results

The analysis of the fate of the NOM indicated that 20% of NOM passed through the SPE column, 59% of NOM was eluted by methanol and present in the final 0.5 mL of eluent. Only 0.4% of NOM was removed during the column rinse step. The final 20% of NOM was believed to

Table 6.2
Effect of NOM on HAA recoveries

HAAs only	HAAs+1 ppm NOM as C	HAAs+5 ppm NOM as C
63 ± 7.6	62 ± 1.4	65 ± 0.25
96 ± 2.3	94 ± 1.0	97 ± 3.2
93 ± 4.4	95 ± 1.2	92 ± 1.2
98 ± 3.3	89 ± 1.6	93 ± 6.9
96 ± 1.4	95 ± 6.3	95 ± 3.5
95 ± 1.6	96 ± 6.1	98 ± 5.2
99 ± 1.5	92 ± 3.3	93 ± 7.5
85 ± 2.5	91 ± 6.5	100 ± 36
74 ± 1.5	77 ± 4.7	78 ± 9.7

have been retained on the SPE column. Although 59% of the NOM was present in the final 0.5 mL of eluent, the recoveries of HAAs in the solution spiked with NOM and the solution without NOM were similar (Table 6.2). Electrophoregrams from the analysis of the sample containing HAAs with 5 mg/L as C NOM and the NOM-only blank sample after SPE preconcentration are shown in Figures 6.2 and 6.3, respectively. The only effect of NOM observed in the electropherograms was an elevated baseline. Furthermore, the few small contaminant peaks in the blank (NOM-only) electropherogram (Figure 6.3) did not interfere with the HAA peaks. Thus, the presence of NOM did not affect the recoveries of HAAs by the SPE process.

STABILITY OF HAAs SORBED ON SPE COLUMNS

Introduction

The use of SPE for sample preconcentration offers two other potential advantages: 1) it is more convenient to load samples onto the SPE columns in the field and then ship only the columns to the analytical laboratories for subsequent HAA elution and analysis and 2) the HAAs may be more resistant to degradation if removed from the aqueous phase for shipment. In order to test the latter hypothesis, the recoveries of HAAs as a function of SPE column storage time were determined.

Materials and Methods

Four samples containing 50 mL of 50 µg/L HAAs and 200 µg/L TBAA were prepared from aqueous HAA stock solutions. The samples were adjusted to pH 2 with sulfuric acid and then passed through four separate SPE columns. The columns were stored at 4°C in the refrigerator and eluted and analyzed on different days.

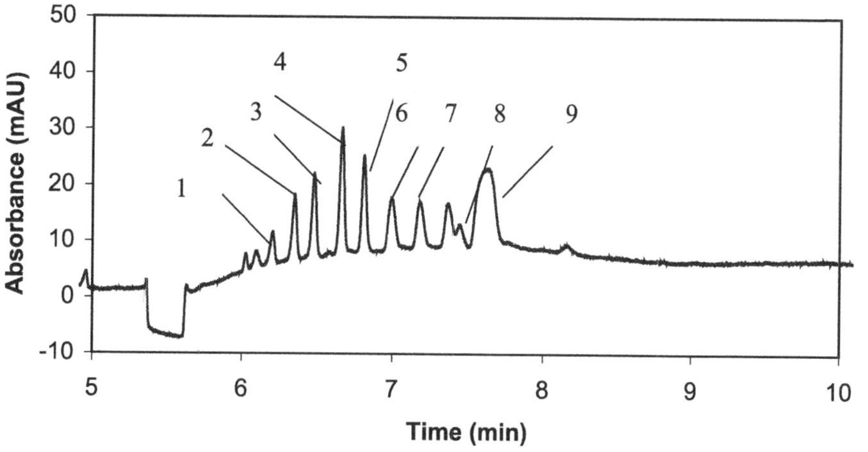

Figure 6.2 CE electropherogram from the analysis of SPE preconcentrated analysis of a 50 mL sample originally containing 50 ppb HAAs +5 mg/L NOM carbon
(Peak identification: 1-CAA, 2-BAA, 3-DCAA, 4-BCAA, 5-DBAA, 6-TCAA, 7-BDCAA, 8-CDBAA, 9-TBAA; Buffer composition: 12.5 mM NaH_2PO_4, 12.5 mM Na_2HPO_4, 1 mM TTAOH, pH = 5.7; Run conditions: Temperature = 25°C, Voltage = –20 kV, Injection = pressure (600 mbar·s), Detection = direct UV at 191 nm; Capillary: bubble cell (75 μm ID × 56 cm effective length))

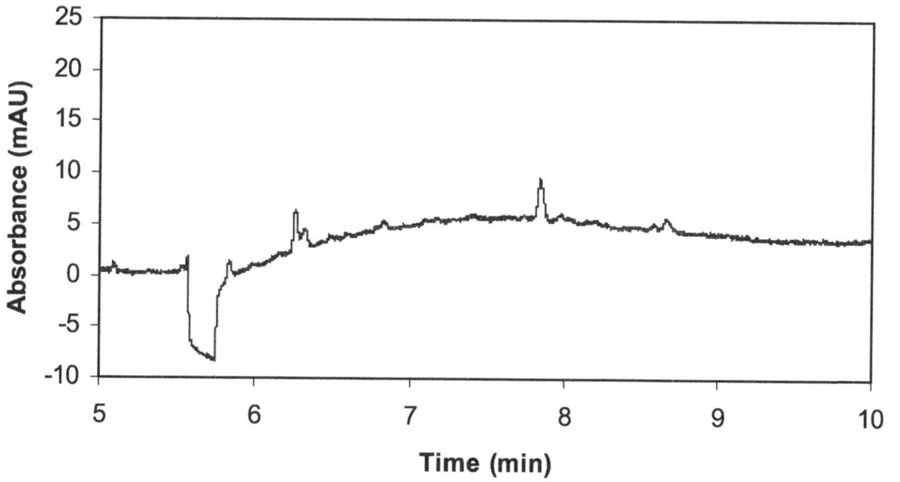

Figure 6.3 CE electropherogram from the SPE preconcentrated analysis of a 50 mL NOM only blank solution
(Buffer composition: 12.5 mM NaH_2PO_4, 12.5 mM Na_2HPO_4, 1 mM TTAOH, pH = 5.7; Run conditions: Temperature = 25°C, Voltage = –20 kV, Injection = pressure (600 mbar·s), Detection = direct UV at 191 nm; Capillary: bubble cell (75 μm ID × 56 cm effective length))

Table 6.3
HAA recoveries as a function of SPE column storage time

	Time (days)			
HAA	0	1	2	5
CAA	0.63	0.61	0.68	0.70
BAA	0.94	0.91	0.96	0.95
DCAA	0.91	0.92	0.79	0.83
BCAA	0.87	0.79	0.85	0.89
DBAA	0.94	0.88	0.90	0.90
TCAA	0.92	0.90	1.00	1.00
BDCAA	0.95	0.94	1.05	0.94
CDBAA	1.01	0.83	0.82	0.95
TBAA	0.84	0.51	0.56	0.40

Results

As shown in Table 6.3, the recovery of TBAA decreased over time while the recoveries of the other HAAs remained relatively stable over the 5 day period. Although the HAAs are predominantly sorbed onto the solid resin, it is possible that TBAA may undergo hydrolysis or decarboxylation (Urbansky 2001, Zhang and Minear 2002) because a small amount of water likely remains in the column after sample processing. Because TBAA is used as the surrogate for evaluating SPE extraction efficiency, this apparent degradation during storage suggests that it may not be feasible to load samples onto the columns in the field and ship the columns back to the laboratory for analysis. More work is needed to verify these TBAA loss results and to investigate whether drying the column prior to storage or other approaches can be used to improve TBAA recoveries.

FINAL METHOD

An overview of the final method is included below.

SPE

The SPE process includes several steps: a) preconditioning of the column with 30 mL of methanol and 10 mL of pH 2 HPLC grade water, b) acidification of the sample to pH 2, c) sample loading at 1 mL/min, d) column rinsing with 0.5 mL of pH 2 HPLC grade water, e) elution of HAAs with 2.3 mL of methanol, f) dilution with 0.2 mL of HPLC grade water, and g) evaporation of methanol prior to analysis by CE.

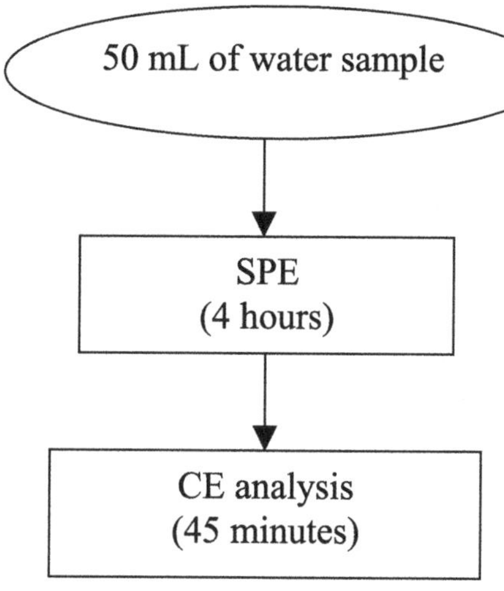

Figure 6.4 Flow chart of proposed SPE-CE method

CE Analysis

The buffer contains 12.5 mM NaH_2PO_4, 12.5 mM Na_2HPO_4 and 1 mM TTAOH at pH=5.7. A bubble cell capillary (75 μm ID × 56 cm effective length) is used for HAA separation and detection. Pressure injection (50 mbar for 12 s) is used to load the samples onto the capillary and the detection wavelength is set at 191 nm.

The flow chart of the final method including approximate times for each step is shown in Figure 6.4. The SPE sample processing is time consuming, but much of this process involves passive waiting time. For example, the time for the sample to pass through the column is approximately 30 minutes and the methanol evaporation step takes more than 2 hours. The CE analysis time includes time for system setup and cleaning of the capillary prior to analysis of the sample.

COMPARISON OF THE SPE-CE METHOD AND AN EXTRACTION-METHYLATION–GC-ECD METHOD

Introduction

Finished water samples obtained from the SPRWS distribution system were analyzed for HAAs by the CE-based method and the results were compared to measurements obtained by an outside laboratory using EPA Method 552.2. The SPRWS system serves 343,000 customers over an area of 114 square miles. The 50-million-gallon-per-day (annual average flow) SPRWS treatment plant receives source water from a chain of impoundment reservoirs predominantly supplied by a diversion from the Mississippi River. The water treatment train consists of lime softening with aluminum sulfate (alum) and ferric chloride coagulation, flocculation, clarification, recarbonation, disinfection with free chlorine followed by ammonia addition for chloramine formation, anthracite-sand filtration, and finally fluoridation. The chlorine dose is 5 mg/L and the free chlorine contact time is approximately 1.9 minutes before the ammonia is added. The water samples were all collected on October 14, 2002 from a single location in the distribution system

Table 6.4
Physical and chemical characteristics of SPRWS plant effluent water (mean values for October 2002)

Parameter	Value
Temperature (°C)	11
Turbidity (NTU)	0.069
Alkalinity-Total (mg/L)	55
pH	8.98
Dissolved Oxygen (mg/L)	11.0
Residual Chlorine (mg/L)	3.4
Total Dissolved Solids (mg/L)	174
Carbonate Hardness (mg/L as $CaCO_3$)	55
Non-Carbonate Hardness (mg/L as $CaCO_3$)	31
Total Hardness (mg/L as $CaCO_3$)	86

with an approximate residence time of 2 to 3 hours. A summary of the finished water characteristics for October 2002 is included as Table 6.4. The total organic carbon (TOC) concentration for the plant effluent on September 16, 2002 was 5.3 mg/L.

Materials and Methods

For this phase of the method testing, a water sample was collected and split, with half of the sample sent to an outside laboratory for HAA analysis by Method 552.2 and the other half brought back to our laboratory for analysis using the SPE-CE method. All glassware used for sample collection and mixing was acid washed, rinsed with DI water, and baked at 550°C for 6 hours. The ambient HAA samples were collected directly from a tap at a pump station into two 250-mL brown glass sample bottles containing 2.5 mL of 10 mg/mL ammonium chloride as preservative. The ambient sample was later spiked with TBAA to a concentration of 200 µg/L in our laboratory for SPE-CE analysis. In addition, two other water samples were collected from the same location. One was spiked with 10 µg/L HAAs and 40 µg/L TBAA and the other was spiked with 50 µg/L HAAs and 200 µg/L TBAA. The samples were spiked with stock solutions containing 10 mg/L HAAs and 40 mg/L TBAA and made to volume with DI water in a 1-L volumetric flask. The HAA-spiked samples were then split into separate brown glass sample bottles containing preservative as described above. The samples were immediately transported by car to our laboratory and the outside laboratory in coolers on ice and were stored in the refrigerator prior to analysis. The aqueous samples were extracted and analyzed within one week of collection. Triplicate SPE extractions and analyses were performed on the ambient and two

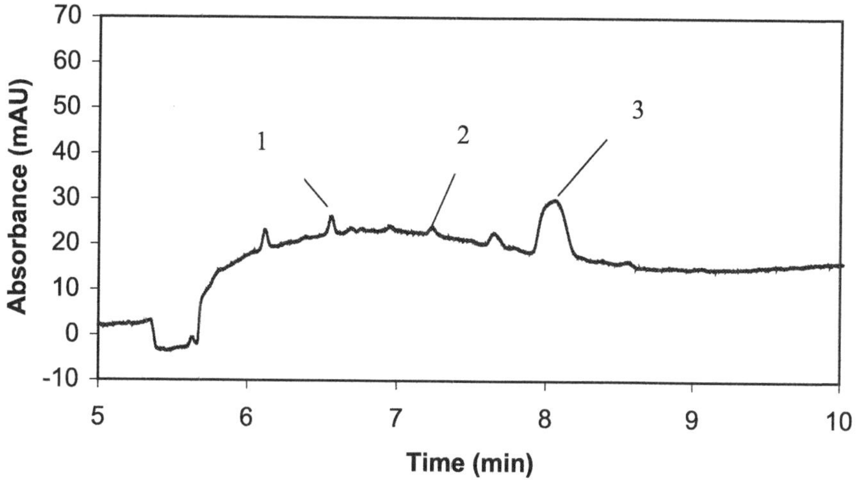

Figure 6.5 CE electropherogram from the analysis of a 50 mL SPRWS water sample (Peak identification: 1-DCAA, 2-TCAA, 3-TBAA; Buffer composition: 12.5 mM NaH_2PO_4, 12.5 mM Na_2HPO_4, 1 mM TTAOH, pH = 5.7; Run conditions: Temperature = 25°C, Voltage = –20 kV, Injection = pressure (600 mbar·s), Detection = direct UV at 191 nm; Capillary: bubble cell (75 μm ID × 56 cm effective length))

spiked water samples in our laboratory using a 50 mL sample volume and one extraction and analysis was performed using a 25 mL sample volume. Only one extraction and analysis per sample was performed by the outside laboratory.

Results and Discussion

Electropherograms from the analysis of the ambient water sample and the spiked water samples are shown in Figures 6.5 to 6.7. A representative electropherogram from the analysis of a 50 mL blank (HPLC grade water) is shown in Figure 6.8. DCAA and TCAA were the only two HAAs detected in the ambient sample by the CE method. The DCAA and TCAA peaks were identified by sample spiking with 2 ppm DCAA or 2 ppm TCAA. However, CAA, BCAA, DCAA, and TCAA were detected by Method 552.2. A comparison of the results from the two methods for each individual HAA and for HAA_5 between the CE-based method and an extraction-methylation-GC-ECD method is shown in Table 6.5, 6.6 and Figures 6.9 and 6.10. The HAA MDLs are reported in Table 6.7. As can be seen, the results obtained in our laboratory were in good agreement with the results obtained by Method 552.2 from an outside laboratory.

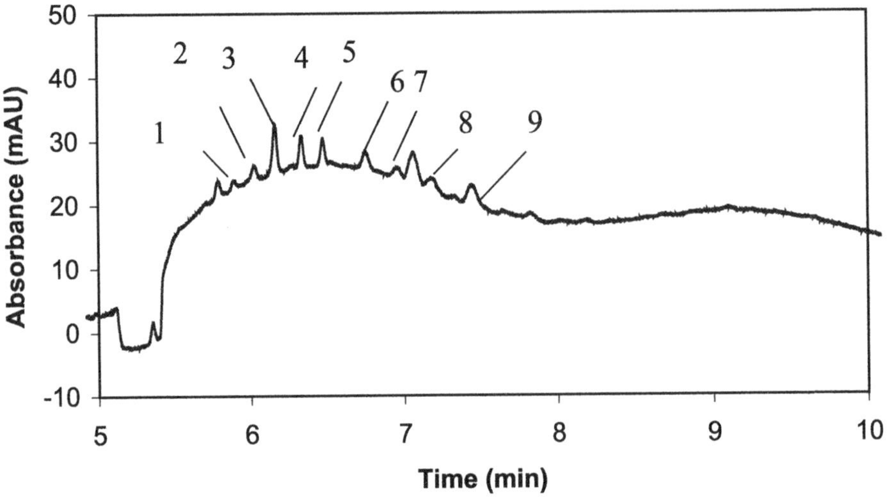

Figure 6.6 CE electropherogram from the analysis of a 50 mL sample of SPRWS water spiked with 10 μg/L HAAs and 40 μg/L TBAA
(Peak identification: 1-CAA, 2-BAA, 3-DCAA, 4-BCAA, 5-DBAA, 6-TCAA, 7-BDCAA, 8-CDBAA, 9-TBAA; Buffer composition: 12.5 mM NaH_2PO_4, 12.5 mM Na_2HPO_4, 1 mM TTAOH, pH = 5.7; Run conditions: Temperature = 25°C, Voltage = –20 kV, Injection = pressure (600 mbar·s), Detection = direct UV at 191 nm; Capillary: bubble cell (75 μm ID × 56 cm effective length))

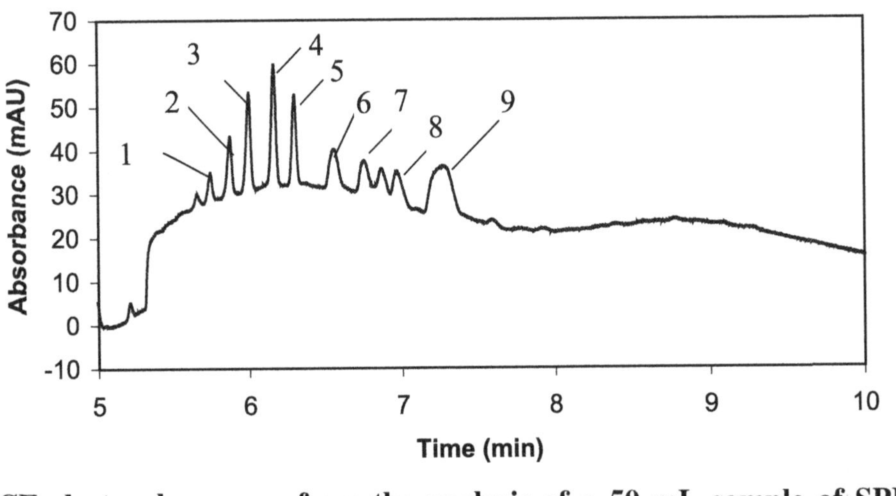

Figure 6.7 CE electropherogram from the analysis of a 50 mL sample of SPRWS water spiked with 50 μg/L HAAs and 200 μg/L TBAA
(Peak identification: 1-CAA, 2-BAA, 3-DCAA, 4-BCAA, 5-DBAA, 6-TCAA, 7-BDCAA, 8-CDBAA, 9-TBAA; Buffer composition: 12.5 mM NaH_2PO_4, 12.5 mM Na_2HPO_4, 1 mM TTAOH, pH = 5.7; Run conditions: Temperature = 25°C, Voltage = –20 kV, Injection = pressure (600 mbar·s), Detection = direct UV at 191 nm; Capillary: bubble cell (75 μm ID × 56 cm effective length))

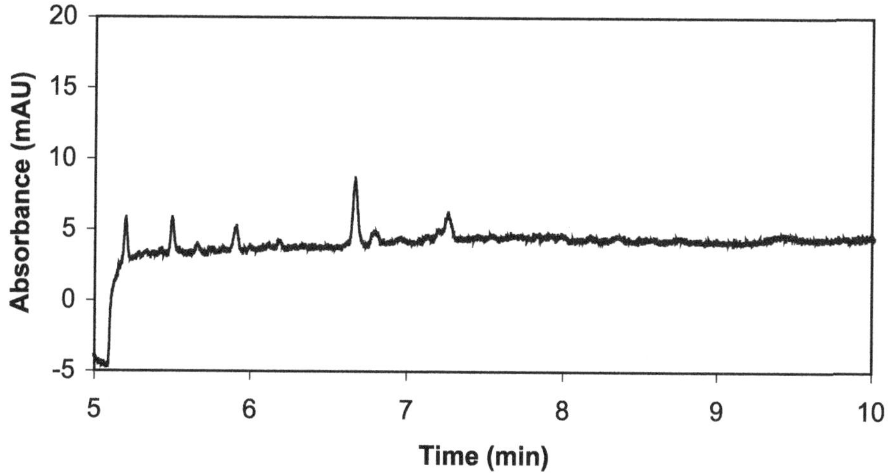

Figure 6.8 CE electropherogram from the analysis of a 50 mL blank sample (Buffer composition: 12.5 mM NaH_2PO_4, 12.5 mM Na_2HPO_4, 1 mM TTAOH, pH = 5.7; Run conditions: Temperature = 25°C, Voltage = –20 kV, Injection = pressure (600 mbar·s), Detection = direct UV at 191 nm; Capillary: bubble cell (75 μm ID × 56 cm effective length))

Table 6.5
Comparison of HAA concentrations obtained by the SPE-CE method and Method 552.2

	CAA (μg/L)		BAA (μg/L)		DCAA (μg/L)		BCAA (μg/L)		DBAA (μg/L)		TCAA (μg/L)	
Sample	552.2	SPE-CE	552.2	SPE-CE	552.2	SPE-CE	552.2	SPE-CE	552.2	SPE-CE	552.2	SPE-CE
Ambient	ND	ND	ND	ND	22	17.5 ± 2.2	2.4	ND	ND	ND	12	9.68 ± 0.96
10 μg/L spike	14	12.6 ± 0.8	11	12.4 ± 0.7	32	27.1 ± 0.7	19	13.9 ± 0.4	12	12.2 ± 0.5	22	17.4 ± 0.2
50 μg/L spike	62	53 ± 0.5	53	56.6 ± 2.4	76	72.5 ± 6	76	72.4 ± 3.8	49	55.5 ± 2.6	60	56.0 ± 1.4

ND = not detected

Table 6.6
Comparison of HAA_5 concentrations obtained by the SPE-CE method and Method 552.2

	HAA_5 (μg/L)	
Sample	Method 552.2	SPE-CE
Ambient	38.1	36.6 ± 3.1
10 μg/L spike	91	95.9 ± 0.69
50 μg/L spike	300	293 ± 10.9

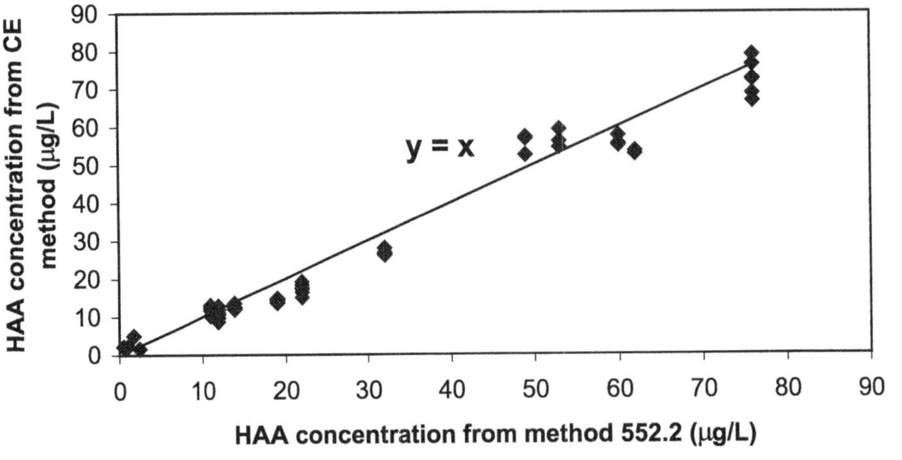

Figure 6.9 Comparison between the CE-based method and EPA method 552.2 for the analysis of ambient and spiked finished water samples

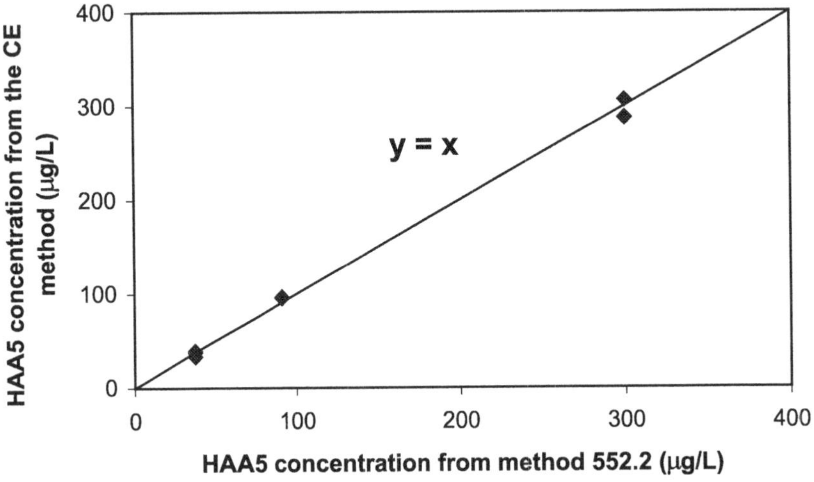

Figure 6.10 Comparison between the CE-based method and EPA method 552.2 for the analysis of HAA_5

PEAK IDENTIFICATION IN CE

Introduction

Peak identification is an important issue in method development. In CE, absolute migration times often vary by several percent or more due to changes in the electroosmotic flow velocity resulting from changes in capillary cleanliness and buffer composition (e.g., pH). This variability in peak migration time can create problems for peak identification. Thus, two approaches for peak identification were considered: 1) use sample spiking with each individual analyte and 2) use relative migration times. Spiking should work effectively, but many more analyses would be required which is undesirable. This section comes up a method based on use of

Table 6.7
Comparison of HAA MDLs for the SPE-CE method and EPA Method 552.2

	MDLs (μg/L)	
HAA	Method 552.2	SPE-CE method
CAA	1.7	5
BAA	1.0	2.2
DCAA	1.5	1.8
BCAA	1.0	1.7
DBAA	0.5	2.2
TCAA	0.5	2
BDCAA	ND	2.1
CDBAA	ND	2.3
TBAA	ND	2.4

ND = not determined

relative migration times, in which the migration times of the analytes are normalized by a marker compound. In these experiments, the marker compound was TBAA, which also serves as the surrogate for the SPE preconcentration process.

Materials and Method

The ambient SPRWS water sample, water sample spiked with 50 μg/L of HAAs and 200 μg/L of TBAA, and water sample spiked with 10 μg/L of HAAs and 40 μg/L of TBAA together with aqueous standards were analyzed using CE. The migration times of the HAA peaks in the water samples and standards were obtained from the electropherograms.

Results and Discussion

As shown in Table 6.8, the absolute migration times of the HAAs can vary considerably with replicate analyses of the same sample. The RSD was as high as 6%, which could result in misidentification of the analyte peaks. However, the maximum RSD for the relative migration times was only 2%. Although the relative migration times (calculated as the migration time of the target compound divided by the migration time of TBAA) of the HAAs in the ambient sample and in the spiked samples were different from the relative migration times of HAAs in aqueous standards, the relative migration times of the HAAs in the ambient sample were similar to those in the spiked samples (Table 6.9). Thus, in order to identify the peaks for real water sample runs, it is recommended to analyze one water sample spiked with all nine HAAs including a high concentration of the surrogate TBAA (e.g., 200 μg/L). Then, the relative migration times calculated from the spiked water samples can be used to identify the actual water samples.

Table 6.8
Comparison of absolute HAA migration times for different water samples

	Absolute migration time of water samples (min.)								
Sample	CAA	BAA	DCAA	BCAA	DBAA	TCAA	BDCAA	CDBAA	TBAA
Standard Solution	6.038 ± 0.083	6.153 ± 0.085	6.267 ± 0.086	6.418 ± 0.087	6.541 ± 0.089	6.738 ± 0.091	6.908 ± 0.090	7.088 ± 0.092	7.295 ± 0.097
Ambient			6.187 ± 0.324			6.772 ± 0.397			
10 μg/L spiked	5.848 ± 0.04	5.979 ± 0.039	6.108 ± 0.042	6.274 ± 0.044	6.416 ± 0.046	6.682 ± 0.050	6.886 ± 0.057	7.109 ± 0.061	7.359 ± 0.067
50 μg/L spiked	6.056 ± 0.210	6.205 ± 0.226	6.346 ± 0.243	6.532 ± 0.266	6.684 ± 0.284	6.965 ± 0.317	7.192 ± 0.344	7.357 ± 0.401	7.746 ± 0.422

Table 6.9
Comparison of relative HAA migration times for different water samples

	Migration time of HAAs relative to TBAA							
Sample	CAA	BAA	DCAA	BCAA	DBAA	TCAA	BDCAA	CDBAA
Standard Solution	0.828 ± 0.003	0.843 ± 0.004	0.859 ± 0.003	0.880 ± 0.004	0.897± 0.004	0.924 ± 0.003	0.947 ± 0.002	0.972 ± 0.001
Ambient			0.826 ± 0.010			0.904 ± 0.005		
10 μg/L spiked	0.795 ± 0.003	0.813 ± 0.003	0.830 ± 0.003	0.853 ± 0.003	0.872 ± 0.003	0.908 ± 0.002	0.936 ± 0.001	0.966 ± 0.001
50 μg/L spiked	0.782 ± 0.016	0.802 ± 0.015	0.820 ± 0.013	0.844 ± 0.012	0.863 ± 0.010	0.899 ± 0.008	0.929 ± 0.006	0.950 ± 0.002

SUMMARY

HAA peak area and peak shape were not affected by the presence of natural organic matter and inorganic anions such as chloride and bromide. Good agreement between the CE method developed in our laboratory and the EPA method 552.2 was observed for analysis of HAAs in a real drinking water sample and spiked drinking water samples. The relative migration time of the HAAs in ambient water samples were equal to those in HAAs spiked samples. Thus, it is recommended to run a spiked water sample first for peak identification purposes.

CHAPTER 7
CONCLUSIONS AND RECOMMENDATIONS

In the HAA analysis method developed in this research project, SPE was used for HAA preconcentration followed by CE analysis. SPE using LiChrolut EN columns was successful at enriching all nine HAAs. The SPE process includes six steps: 1) preconditioning of the column with 30 mL of methanol and 10 mL of pH 2 HPLC grade water, 2) acidification of the sample to pH 2, 3) sample loading at 1 mL/min, 4) column rinsing with 0.5 mL of pH 2 HPLC grade water, 5) elution of HAAs with 2.3 mL of methanol, 6) dilution with 0.2 mL of HPLC grade water and 7) evaporation of methanol prior to analysis. TBAA (200 μg/L) is the recommended surrogate for determining HAA extraction efficiencies. For the analysis of CAA, it is recommended that only 25 mL sample be processed in order to obtain acceptable recoveries. However, 50 mL of water sample is recommended for the analysis of the other HAAs. The recoveries were in the range of 75% to 99% for all HAAs except CAA (63%) when 50 mL of sample was passed through the column and the recovery for CAA increased to 80% when 25 mL of sample was used. If the 63% CAA recovery is deemed acceptable by the analyst, then only one extraction and analysis per water sample would be required.

The CE analysis conditions are as follows. The buffer contains 12.5 mM NaH_2PO_4, 12.5 mM Na_2HPO_4 and 1 mM TTAOH at pH=5.7. A bubble cell capillary (75 μm ID × 56 cm effective length) is used for HAA separation and detection. Pressure injection (50 mbar for 12 s) is used to load the samples onto the capillary and the detection wavelength is set at 191 nm.

The SPE-CE method can be used to analyze for all HAAs except CDBAA (comigrated with a contamination peak) and TBAA (used as surrogate). The overall MDLs including SPE preconcentration and analysis by CE ranged from 1.7 to 2.5 μg/L for all HAAs except CAA (5 μg/L). It takes approximately 5 hours to process each sample, however, by the use of a multi-port vacuum manifold, it is possible to process several samples at the same time.

Two methods can be used to identify HAA peaks in the electropherogram. The first method involves standard addition. The second method is to analyze a water sample spiked with HAAs and the surrogate TBAA. Then, use the relative HAA migration times calculated from the spiked water sample to identify the HAAs in the actual water samples.

REFERENCES

Ahrer, W., and W. Buchberger. 1999. Determination of Haloacetic Acids by the Combination of Non-Aqueous Capillary Electrophoresis and Mass Spectrometry. *Fresenius J. Anal. Chem.*, 365(7):604–609.

Altria, K.D. 1999. Optimization of Sensitivity in Capillary Electrophoresis. *LC-GC: Magazine of Liq. Gas Chromatogr.*, 17(1).

APHA, AWWA, and WEF (American Public Health Association, American Water Works Association, and Water Environment Federation). 1995. *Standard Methods for the Examination of Water and Wastewater.* 19th ed. Washington, D.C.: APHA.

APHA, AWWA, and WEF (American Public Health Association, American Water Works Association, and Water Environment Federation). 1998. *Standard Methods for the Examination of Water and Wastewater.* 20th ed. Washington, D.C.: APHA.

Benanou, D., F. Acobas, and P. Sztajnbok. 1998. Analysis of Haloacetic Acids in Water By a Novel Technique—Simultaneous Extraction-Derivatization. *Water Res.*, 32(9):2798–2806.

Berg, M., S.R. Müller, J. Muhlemann, A. Wiedmer, and R.P. Schwarzenbach. 2000. Concentrations and Mass Fluxes of Chloroacetic Acids and Trifluoroacetic Acid in Rain and Natural Waters in Switzerland. *Environ. Sci. Technol.*, 34(13):2675–2683.

Carrero, H., and J.F. Rusling. 1999. Analysis of Haloacetic Acid Mixtures by HPLC Using an Electrochemical Detector Coated With a Surfactant-Nafion Film. *Talanta*, 48(3):711–718.

Ells, B., D.A. Barnett, R.W. Purves, and R. Guevremont. 2000. Detection of Nine Chlorinated and Brominated Haloacetic Acids at Part-per-Trillion Levels Using ESI-FAIMS-MS. *Anal. Chem.*, 72(19):4555–4559.

Foret, F., E. Szoko, and B.L. Larger. 1992. On-Column Transient and Coupled Column Isotachophoretic Preconcentration of Protein Samples in Capillary Zone Electrophoresis. *J. Chromatogr.*, 608:3–12.

Galceran, M.T., L. Puignou, and M. Diez. 1996. Comparison of Different Electroosmotic Flow Modifiers in the Analysis of Inorganic Anions By Capillary Electrophoresis. *J. Chromatogr. A*, 732(1):167–174.

Glynn, J.R., B.M. Belongia, R.G. Arnold, K.L. Ogden, and J.C. Baygents. 1998. Capillary Electrophoresis Measurements of Electrophoretic Mobility for Colloidal Particles of Biological Interest. *Appl. Environ. Microbiol.*, 64(7):2572–2577.

Heller-Grossman, L., J. Manka, B. Limonirelis, and M. Rebhun. 1993. Formation and Distribution of Haloacetic Acids, THM and TOX in Chlorination of Bromide-Rich Lake Water. *Water Res.*, 27(8):1323–1331.

Kampioti, A.A., and E.G. Stephanou. 1999. Simultaneous Determination of Halogenated Neutral and Acidic Disinfection By-Products in Drinking Water by Closed-Loop Stripping Extraction and Capillary Gas Chromatography. *J. Chromatogr. A*, 857(1–2):217–229.

Kaniansky, D., F. Ivanyl, and F.I. Onuska. 1990. On-Line Coupling of Capillary Isotachophoresis With Capillary Zone Electrophoresis. *J. Chromatogr.*, 498:191–204.

Kaplan, L.A. 2000. Comparison of Three TOC Methodologies. *Jour. AWWA*, 92(4):149–156.

Kim, D.H., J.O. Choi, M. Kim, and D.W. Lee. 2001. Determination of Haloacetic Acids in Tap Water by Capillary Electrophoresis With Direct UV Detection. *J. Liq. Chromatogr. Related Technol.*, 24(1):47–55.

Krasner, S.W., M.J. McGuire, J.G. Jacangelo, N.L. Patania, K.M. Reagan, and E.M. Aieta. 1989. Occurrence of Disinfection By-Products in U.S. Drinking Water. *Jour. AWWA*, 81:41–53.

Krasner, S.W., S. Pastor, R. Chinn, M.J. Scilimenti, H.S. Weinberg, S.D. Richardson, and A.D. Truston. 2000. The Occurrence of a New Generation of DBPs (Beyond the ICR). In *Proc. of the AWWA Water Quality Technology Conference*. Denver, Colo.: AWWA.

Krivankova, L., and P. Bocek. 1998. Capillary Isotachophoresis. In *High Performance Capillary Electrophoresis Theory, Techniques and Application*. Edited by G.K. Morteza. New York: John Wiley & Sons.

Loos, R., and D. Barcelo. 2001. Determination of Haloacetic Acids in Aqueous Environments by Solid-Phase Extraction Followed by Ion-Pair Liquid Chromatography-Electrospray Ionization Mass Spectrometric Detection. *J. Chromatogr. A*, 938(1–2):45–55.

Loos, R., and R. Niessner. 1998. Analysis of Aromatic Sulfonates in Water By Solid-Phase Extraction and Capillary Electrophoresis. *J. Chromatogr.*, 822(2):291–303.

Loos, R., and R. Niessner. 1999. Analysis of Atrazine, Terbutylazine and Their N-dealkylated Chloro and Hydroxy Metabolites by Solid-Phase Extraction and Gas Chromatography Mass Spectrometry and Capillary Electrophoresis-Ultraviolet Detection. *J. Chromatogr.*, 835(1–2):217–229.

Lopez-Avila, V. 2000. *Development of an Experimental Procedure for Haloacetic Acids Using Ion Chromatography.* Denver, Colo.: AwwaRF and AWWA.

Magnuson, M.L., and C.A. Kelty. 2000. Microextraction of Nine Haloacetic Acids in Drinking Water at Microgram per Liter Levels With Electrospray-Mass Spectrometry of Stable Association Complexes. *Anal. Chem.*, 72(10):2308–2312.

Martinez, D., F. Borrull, and M. Calull. 1998a. Comparative Study of a Solid-Phase Extraction System Coupled to Capillary Electrophoresis in the Determination of Haloacetic Compounds in Tap Water. *J. Chromatogr. A*, 827(1):105–112.

Martinez, D., F. Borrull, and M. Calull. 1999. Evaluation of Different Electrolyte Systems and On-Line Preconcentrations for the Analysis of Haloacetic Acids by Capillary Zone Electrophoresis. *J. Chromatogr. A*, 835:187–196.

Martinez, D., F. Borrull, M. Calull, J. Ruana, and A. Colom. 1998b. Application of Solid-Phase Extraction Membrane Disks in the Determination of Haloacetic Acids in Water by Gas Chromatography Mass Spectrometry. *Chromatographia*, 48(11–12):811–816.

Martinez, D., J. Farre, F. Borrull, M. Calull, J. Ruana, and A. Colom. 1998c. Capillary Zone Electrophoresis With Indirect UV Detection of Haloacetic Acids in Water. *J. Chromatogr. A*, 808(1–2):229–236.

Müller, S.R., H.R. Zweifel, D.J. Kinnison, J.A. Jacobsen, M.A. Meier, M.M. Ulrich, and R.P. Schwarzenbach. 1996. Occurrence, Sources, and Fate of Trichloroacetic Acid in Swiss Waters. *Environ. Toxicol. Chem.*, 15(9):1470–1478.

National Academy of Science Safe Drinking Water Committee. 1980. *Drinking Water and Health.* Washington, D.C.: National Academy Press.

Oxspring, D.A., W.F. Smyth, and R. Marchant. 1995. Comparison of Reversed-Polarity Capillary Electrophoresis and Adsorptive Stripping Voltammetry for the Detection and Determination of Reactive Textile Dyes. *Analyst*, 120(7):1995–2000.

Rebhun, M., L. Heller-Grossman, and J. Manka. 1997. Formation of Disinfection Byproducts During Chlorination of Secondary Effluent and Renovated Water. *Water Environ. Res.*, 69(6):1154–1162.

Roehl, R., R. Slingsby, N. Avdalovic, and P.E. Jackson. 2002. Applications of Ion Chromatography With Electrospray Mass Spectrometric Detection to the Determination of Environmental Contaminants in Water. *J. Chromatogr. A*, 956(1–2):245–254.

Sarrion, M.N., F.J. Santos, and M.T. Galceran. 2000. In Situ Derivatization/Solid Phase Microextraction for the Determination of Haloacetic Acids in Water. *Anal. Chem.*, 72(20):4865–4873.

Sarzanini, C., M.C. Bruzzoniti, and E. Mentasti. 1999. Preconcentration and Separation of Haloacetic Acids by Ion Chromatography. *J. Chromatogr. A*, 850(1–2):197–211.

Scott, B.F., D. Mactavish, C. Spencer, W.M.J. Strachan, and D.C.G. Muir. 2000. Haloacetic Acids in Canadian Lake Waters and Precipitation. *Environ. Sci. Technol.*, 34(20):4266–4272.

Serjeant, E.P., and B. Dempsey. 1979. *Ionisation Constants of Organic Acids in Aqueous Solution.* New York: Pergamon Press.

Smyth, W.F., G.B. Harland, S. McClean, G. McGrath, and D. Oxspring. 1997. Effect of On-Capillary Large Volume Sample Stacking On Limits of Detection in the Capillary Zone Electrophoretic Determination of Selected Drugs, Dyes and Metal Chelates. *J. Chromatogr. A*, 772(1–2):161–169.

Urbansky, E.T. 2001. The Fate of the Haloacetates in Drinking Water: Chemical Kinetics in Aqueous Solution. *Chem. Rev.*, 101(11):3233–3243.

Urbansky, E.T., and M.L. Magnuson. 2002. Analyzing Drinking Water for Disinfection Byproducts. *Anal. Chem.*, 74(9):1.

USEPA (U.S. Environmental Protection Agency). 1992. *Method 552.1: Determination of Haloacetic Acids and Dalapon in Drinking Water by Ion-Exchange Liquid-Solid Extraction and Gas Chromatography With an Electron Capture Detector.* Cincinnati, Ohio: Environmental Monitoring and System Laboratory.

USEPA (U.S. Environmental Protection Agency). 1995. *Method 552.2: Determination of Haloacetic Acids and Dalapon in Drinking Water by Liquid-Liquid Extraction, Derivatization and Gas Chromatography With Electron Capture Detection.* Cincinnati, Ohio: Environmental Monitoring and System Laboratory.

USEPA (U.S. Environmental Protection Agency). 2000. *Stage II Microbial/Disinfection Byproducts (M/DBP) Agreement in Principle.*

Windholz, M., S. Budavari, R.F. Blumetti, and E.S. Otterbein. 1983. *Merck Index.* Rahway, N.J.: Merck.

Xie, Y.F. 2001. Analyzing Haloacetic Acids Using Gas Chromatography/Mass Spectrometry. *Water Res.*, 35(6):1599–1602.

Xie, Y.F., D.A. Reckhow, and D.C. Springborg. 1998. Analyzing HAAs and Ketoacids Without Diazomethane. *Jour. AWWA*, 90(4):131–138.

Xie, Y.F., and P. Romano. 1997. Application of Capillary Electrophoresis for Drinking Water Analysis. In *Proc. of the AWWA Water Quality Technology Conference.* Denver, Colo.: AWWA.

Xie, Y.F., H. Zhou, and J.P. Romano. 1999. Development of a Capillary Electrophoresis Method for Haloacetic Acids. *American Chemical Society National Meeting, Division of Environmental Chemistry*, 39:259–261.

Zhang, X.R., and R.A. Minear. 2002. Decomposition of Trihaloacetic Acids and Formation of the Corresponding Trihalomethanes in Drinking Water. *Water Res.*, 36(14):3665–3673.

ABBREVIATIONS

APHA	American Public Health Association
AWWA	American Water Works Association
AwwaRF	Awwa Research Foundation
BAA	bromoacetic acid
BCAA	bromochloroacetic acid
BDCAA	bromodichloroacetic acid
°C	degrees Celsius
CAA	chloroacetic acid
CDBAA	chlorodibromoacetic acid
CE	capillary electrophoresis
CITP	capillary isotachophoresis
cm	centimeter
CTABr	cetyltrimethylammonium bromide
CTACl	cetyltrimethylammonium chloride
CTAOH	cetyltrimethylammonium hydroxide
CZE	capillary zone electrophoresis
DBAA	dibromoacetic acid
DBP	disinfection by-product
DCAA	dichloroacetic acid
D/DBP	disinfectants/disinfection by-product
DETA	diethylenetriamine
DI	distilled and deionized
DOC	dissolved organic carbon
EM	electrophoretic mobility
EOF	electroosmotic flow
EPA	Environmental Protection Agency
ESI-FAIMS-MS	electrospray ionization, high-field asymmetric waveform ion mobility spectrometry mass spectrometry
g	gram
GC-ECD	gas chromatography with electron capture detection
GC-ITMS	gas chromatography-ion trap mass spectrometry
GC-MS	gas chromatography-mass spectrometry

HAA	haloacetic acid
HDMBr	hexadimethrine bromide
HDMOH	hexadimethrine hydroxide
HOBr	hypobromous acid
HOCl	hypochlorous acid
HPLC	high-performance liquid chromatography
ID	inner diameter
IDL	instrument detection limit
ITP	isotachophoresis
kV	kilovolt
L	liter
m	meter
M	molar
mAU	milliabsorbance unit
mbar	millibar
MCL	maximum contaminant level
MDL	method detection limit
mg	milligram
min	minute
mL	milliliter
mm	millimeter
mM	millimolar
MNNG	1-methyl-3-nitro-1-nitrosoguanidine
MTBE	methyl-*tert*-butyl-ether
MW	molecular weight
µg	microgram
µm	micrometer
µM	micromolar
NDC	2,6-naphthalenedicarboxylic acid
nm	nanometer
NOM	natural organic matter
RSD	relative standard deviation
s	second
SPE	solid phase extraction
SPME	solid phase microextraction
SPRWS	St. Paul Regional Water Services

TBAA	tribromoacetic acid
TCAA	trichloroacetic acid
TFAA	trifluoroacetic acid
THM	trihalomethane
TOC	total organic carbon
TTABr	tetradecyltrimethylammonium bromide
TTAOH	tetradecyltrimethylammonium hydroxide
USEPA	United States Environmental Protection Agency
UV	ultraviolet
V	volt
WEF	Water Environment Federation

IWA Publishing's authorised EU representative for General Product Safety Regulations is Diane D'Arras, 15 rue Duret, 75116 Paris, France, e-mail: safety@iwap.co.uk.

Printed and bound by CPI Group (UK) Ltd, Croydon, CR0 4YY
17/09/2026
02218810-0011